The New York Times
Circuits

EDITED BY
HENRY FOUNTAIN

INTRODUCTION BY
ANDY ROONEY

ILLUSTRATIONS BY

Mika Gröndahl, Frank O'Connell,

and other graphic artists of *The New York Times*

ST. MARTIN'S PRESS   ❧   NEW YORK

The New York Times

Circuits

How Electronic Things Work

CANCEL

www.stmartins.com

Book design by James Sinclair

Library of Congress Cataloging-in-Publication Data

The New York Times Circuits : how electronic things work / edited by Henry Fountain—
1st ed.
    p. cm.
  ISBN 0-312-28439-X
  1. Electronic apparatus and appliances. 2. Electric circuits. I. Title: Circuits. II. Fountain, Henry.

TK7870 N497 2001
621.381—dc21

2001042400

First Edition: November 2001

10 9 8 7 6 5 4 3 2 1

CONTENTS

ACKNOWLEDGMENTS

Henry Fountain

Back in 1997 and 1998, when *The New York Times* was planning Circuits, its new weekly section on personal technology, the editors realized early on that there would be some explaining to do. You can't write about the effects of technology without taking some time to describe how technology works, so from the beginning the plan was to devote space in the section to the nuts and bolts (if the reader will forgive a metaphor from the Industrial Age) of computers, chips and the Internet. Those first efforts were called Fundamentals, and together they formed a kind of *Michelin Guide* for that strange technological joyride we were all on.

As Circuits (and its readers) grew more comfortable with technology, the focus of the section's how-it-works efforts shifted to electronic devices and gadgets. Some of these are things we use every day; others are the tools of business or transportation or sports or entertainment. And still others are high-tech instruments that are used by scientists to help understand and explain the natural world.

In producing this book, we have included both the earlier Fundamentals and the more recent gadget-oriented efforts. The result, we think, represents an astonishing amount of information. But we also know that this book really only scratches the surface; we hope it will inspire readers to learn even more about the technology that surrounds us.

More than most, this book is a collaboration between writers, editors and artists. Thanks, first and foremost, to the writers and graphic artists whose work is featured here, but particularly to Mika Gröndahl and Frank O'Connell, who created the bulk of the original graphics that appeared in the newspaper and then painstakingly recreated them for the book.

Thanks also to Jim Gorman, the first editor of Circuits, and Kevin McKenna, the current editor, and to other editors who have contributed so much to the evolution of the section: Jeanne Pinder, Bruce Headlam, Karen Freeman, Sandy Harvin and Clark Line.

This book was the brainchild of Mike Levitas, director of book development at *The Times*, and without his guidance and encouragement it never would have been published. Thanks also to Bryan Cholfin of St. Martin's Press, to Bedel Saget and Joe Ward for the loan of the laptop, and to Paul Jean for a very special download. And special appreciation to Savannah Walker for her extreme patience, and to Walker Fountain for his extreme enthusiasm (and, of course, for the light-up sneakers).

By Andy Rooney

It might not seem so to the readers of Circuits, who are regularly regaled in picture and print with new devices related to the developments in a wide range of computing and communications technology, that the age of invention has ended. It has. We have just recently completed the invention of everything.

The things we use to assist us in living by controlling the forces of nature will continue to be changed and improved, but civilization will never come through another period of invention that has improved life on Earth for mankind so much as the one we've come through in the last 125 years.

Some centuries produce more of one thing than another — the Renaissance period, for example, produced extraordinary art. The period civilization has come through since the late 1800s should be known by future historians as The Age of Invention. Progress is oblivious to our method for designating periods of time so The Age isn't tidily contained within the precise years of the 20th century. Modern inventions started coming before 1900 and may have all but stopped coming before 2000. Edison first demonstrated his incandescent lamp in 1879 but a great many of the inventions that dominate our civilization came early in the 20th century.

What improvement in our lives brought about by applied science or by mechanical aids that overcame darkness, silence, inertia, gravity, heat and cold are we going to get in the next 125 years that are comparable in significance to the light bulb, the telephone, the gasoline engine, automobiles, airplanes, radio, television, air conditioning and the computer?

Computers run by electricity (I don't want anyone quibbling about the abacus) were one of the last inventions. Technicians will be busy for the next millennium finding ways to make them do more and faster. They will, however, still be computers.

All the things we use to assist us in living will continue to be made in different shapes, smaller, quieter and more efficient, but the basic patent will obtain. We want almost everything smaller, hand-held, portable and pocket-sized, but there are practical limits to the extent to which we can take miniaturization. Our incessant pursuit of smaller can't continue forever. There is an ideal size for everything and in some cases we have already passed that ideal in diminution. If it became possible to put the Bible on the head of a pin, pins with the Bible on their heads would not be installed in hotel room drawers because no guest would be able to read

them. Similarly, there's no sense making a button, switch or key on a piece of equipment so small that the average finger can't push it without accidentally exerting some of the force on the button next to it.

The machinery we've invented increases the human ability to do work by multiples of tens of thousands. Workers no longer carry the stones by hand to their niche in the pyramid. No one has to smear the glue on the cardboard flap of each filled cereal box with a brush dipped in a glue pot.

It's getting so easy to produce things in large numbers quickly that manufacturing facilities have often overtaken and passed the company's ability to sell what they make. More money is being spent on selling some products than on making them. The package, being part of the sales pitch, costs more than what's in it.

Inventors these days aren't even trying to come up with anything really new. They're all employed by industry making little changes in existing inventions so we'll feel compelled to buy the new model. My 1999 model car is a great improvement over my first car, a used 1935 Dodge an uncle gave me when I was 17, but the car I drive today is not a new invention. Those two cars were built 64 years apart, but they both had four wheels and an engine. By pressing the accelerator pedal with your foot, you could make either of them go faster than it's legal to drive. You don't have to wind down the windows by hand in the new model and a light goes on over the mirror so a woman can apply lipstick without missing a light, but it is otherwise a lot the same — although $25,000 more costly. It is not a new invention.

The radio reached the end of its potential for being technologically developed soon after the heterodyne receiver replaced the crystal set. The Wright Brothers mastered flight in 1903. Television has a way to go in the development of digital technology and systems for delivery but we have pictures that course through the atmosphere and silently enter our homes. That isn't going to be changed by a new shape of the box in our living room.

The one thing that would make it necessary to reinvent invention would be a war that wiped out all but scattered remnants of our civilization. If that happens—and it could happen—we'd have to start over. How many thousands of years would it take for another Edison to come along to invent the light bulb? How long would we have to wait for Diet Coke?

In the future, amazement at our own progress will be reserved for what science, working with technology, accomplishes with the human genome. Together, they're crowding God. We have come through The Age of Invention and entered the period in which biotechnology will dominate change.

The resistance to genetic research and development by government or religious leaders trying to discourage or prevent it will not prevail. Anytime we have learned how to do something, we have done it whether it was good for us or not. If they can produce a human with the body of Arnold Schwarzenegger and the brain of Albert Einstein in the laboratory, they will do it. The danger for us all is that it will turn out to be someone with the body of Einstein and the brain of Schwarzenegger.

Modern inventors, deprived of so much by so many brilliant innovators who preceded them, nonetheless applied for 160,000 patents last year. The great majority of the patents issued were for gadgets or inconsequential modifications of existing inventions.

When someone hard up for conversation at a dinner party asked what my best gadget was, my first inclination was to dismiss it as a foolish question. But on further consideration I concluded that my best gadget is the automatic garage door opener. It is clipped to the underside of the sun visor on the passenger's side of my car. The sun visor is good, too, and more essential. I think of the word "gadget" to mean some relatively new, nonessential item that is closer to being a toy than a tool. This disqualifies the sun visor as a gadget on several counts.

Many of the products available to those interested in modern technology are on the ragged border between gadget and tool. The computer is a tool; the mouse, in my opinion (with which a great many experts disagree), is a gadget that will disappear into that great storage box in the attic together with the roller skate key. It is, however, a mistake to dismiss gadgets because that means ignoring the importance of fun. A great amount of the time spent on computers by educated people is closer to fun than work. It is this aspect of electronic devices, almost as much as their practical application to a lot of jobs, that provides a major market for anything related to them. The editors of this book presumably hope this book will be a part of that market.

An educated person is someone who has been exposed to a great deal of information and has retained a large part of it. Because most information is worthless, we need vast amounts of it and some way to sort through it quickly to extract its gold. While it is not clear whether the capacity of the human brain has remained constant over the centuries or whether it has gradually increased to enable it to comprise the vast amount of new information available to it, there is no question that computers and the storage devices that go with them have vastly increased our ability to remember and instantly recall most of what is known of the history of mankind.

Systems for the storage and retrieval of information so important to history and education are diverse and already plentiful but they will continue to proliferate.

While there is more advanced technology to be considered in the pages of Circuits than the best brains among us have time to contemplate, it is possible for persons besieged by information to console themselves over what appears to be an inadequate memory with the hope that the human brain has a finite capacity for storing facts and that every fact added after the brain has been filled forces out some previously stored piece of information of comparable size or complexity.

Just as the computer is a tool that is in almost every case superior to the mathematical computations being made on it, the literary quality of the e-mail messages being conveyed is dismally poor. The prose written by college students and business people often borders on illiteracy. The great number of devices being produced or developed for storing things are more sophisticated than most of the material being stored. No storage system has yet been developed with a quality sensor that eliminates the dross. It seems unlikely that one will be forthcoming.

In addition to the importance of computers and retrieval systems to education, saving the history of one generation has always been acknowledged as vital to the success of the next. So far, hard drive, floppy disk, microfiche and digital advocates have prevailed over the archivists who want to preserve the real thing, not a picture of it.

In spite of all the storage systems known to or developed by him, Bill Gates recently acquired the incomparable Bettmann Archive collection of 12 million photographs and chose to have it stored in the shaft of a deserted, inaccessible lime mine in Pennsylvania. So much for high-tech storage and retrieval.

The Dutch did the same thing 60 years ago when they hid Rembrandt's "Night Watch" from the Germans during WW II in an air-conditioned room in the nether reaches of a small mountain near Liège whose labyrinthine passages, hollowed out by marl gatherers, were being used for growing mushrooms in the dark

Those who believe the only true history of a culture resides in actual artifacts are at war with the progressive developers of systems for saving things in some other form. In relatively recent years libraries running out of storage space for daily newspapers and worried about decomposition have turned to photography and are storing negatives of newspapers in one one-hundredth the space the newspapers themselves took up, where they can be located and reviewed in one-tenth the time on any one of a dozen systems.

The applicability of information storage and retrieval to education is inestimable. The major sources of knowledge are more widely available to everyone than they have ever been. I've never heard whether the traditional four-year college course was invented or just got that way, but it seems probable that, like so many things, it will make sense to shorten a college education as they develop better, faster ways to inculcate knowledge. That is, unless colleges drag their administrative feet under the threat of losing their customers a year early. A three-year college education, comparable to what once took four, seems possible. As a grandparent, I'm thinking of $40,000 now.

It seems likely there are limits to the educability of any of us, but it is just as certain that few of us ever reaches his or her maximum capacity for it because of the clumsy process by which we take it in. It would be a great step forward in education if someone developed a device making it possible for information to be fed directly into the brain through electrodes affixed to the temples. A college wouldn't need a football team or cheerleaders.

Creating the technology to remove unwanted material from the brain in the same manner might be more difficult.

Whatever conclusions future generations come to about the progress of intellect, we can hardly deny that technology has enabled us to live lives that are not only longer but fuller. The life expectancy in 1900 was around 47 years. Today it is about 77 at birth: and if a person reaches the age of 80, he is statistically destined to live to be 90.

Our lives are not only longer in years but electronic devices have put us in touch with our world. The exchange of information, ideas and even trivia has lengthened our lives in another dimension through the great variety of interests with which we are able to fill them. Einstein is lurking here somewhere.

An 80-year-old American has put more into his life in one of his years than his forebears did in all 41 of theirs. There were five million Americans when Benjamin Franklin went to Paris in 1776 and records indicate that fewer than 1,000 of them traveled to Europe. Last year, three million Americans went to France. It is not unusual for a successful American to visit half the countries on Earth, to crisscross his own land a

dozen times with side trips to the Caribbean. Their circle of friends, no longer circumscribed by the boundaries imposed by travel behind a horse-drawn cart or even a jet aircraft, might extend to the ends of the Earth on the Internet

At home, on evenings once filled with darkness and later with the light of candles or kerosene lamps that made reading possible but difficult, people can now do at any hour the things they once did only in daylight because of Thomas Edison's incandescent bulb. They can work, read, or talk over wires with friends, parents or children. They can watch the world go by, be informed, entertained or disgusted by television. At its worst, television is an improvement over what the caveman had to help him idle away the hours from dusk to dawn.

We are not only filling our lives with more, we are seeing changes month to month that used to be decade to decade. Change itself is changing faster than it once did. Our methods for making things happen are being improved more in a year than the pre-Industrial Revolution societies changed things in a century. The modern laptop computer is not only a technological marvel of incredible complexity, it is outdated before the customer who bought it gets it home and out of the box it came in.

The realization that the computer, an aid to the brain, is as important as the mechanical devices that took the load off our backs, is relatively recent. For most of our time here on Earth, we have been more interested in amusement and culture than in technology. Technology and culture are cool to each other. They keep their distance.

The Times's coverage of culture, art and entertainment preceded its interest in technology by many years. As a writer I am a reluctant witness, but if forced to hang by my thumbs until I confessed, I would quickly admit that I believe technology and science are more important to our lives than culture. Without culture it would be a gray world, but technology would survive. Our continuing existence depends, not on music, art, or fiction, but on our ability to outwit nature. It might, of course, be better to expire with culture than endure without it.

If science and technology prevail and the nincompoops are defeated by the indubitable evidence of the progress they provide, there is hope for a good life for all those following us for generations to come. If, on the other hand, they are superseded by those who prefer to hope and pray, wage war and buy lottery tickets, we will be ancestors to no one.

———

Andy Rooney is essayist for 60 Minutes *and a columnist for Tribune Media Services.*

Home

Getting the Picture

Many digital cameras use sensors called charge-coupled devices to convert light into electrical signals. But they need other electronics to change the analog output of the C.C.D. to a digital form that can be stored, transferred and manipulated.

INSIDE A DIGITAL CAMERA

Light passes through the lens, which focuses it on the surface of the charge-coupled device. In the C.C.D., light energy is converted into electrical signals. Good-quality digital cameras have C.C.D.'s with several million pixels.

SENSING THE LIGHT

When the shutter opens, light falls on the C.C.D., and electrons are released in the semiconductive silicon layer. The more intense the light, the more electrons are released and the bigger the resulting charge. These charges are eventually converted into a digital signal.

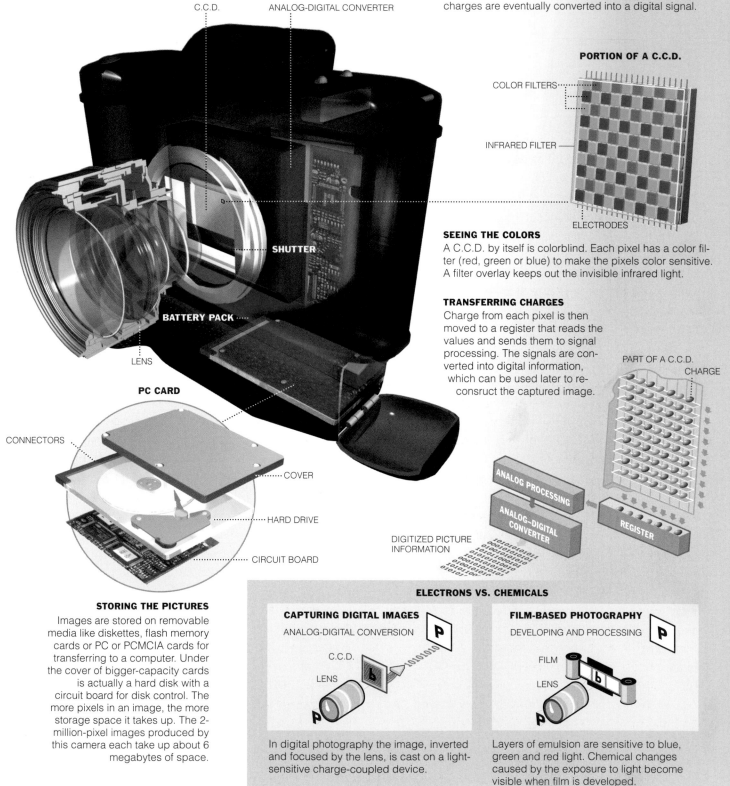

C.C.D.   ANALOG-DIGITAL CONVERTER

SHUTTER

BATTERY PACK

LENS

PC CARD

CONNECTORS

COVER

HARD DRIVE

CIRCUIT BOARD

PORTION OF A C.C.D.

COLOR FILTERS

INFRARED FILTER

ELECTRODES

SEEING THE COLORS

A C.C.D. by itself is colorblind. Each pixel has a color filter (red, green or blue) to make the pixels color sensitive. A filter overlay keeps out the invisible infrared light.

TRANSFERRING CHARGES

Charge from each pixel is then moved to a register that reads the values and sends them to signal processing. The signals are converted into digital information, which can be used later to reconsruct the captured image.

PART OF A C.C.D.

CHARGE

ANALOG PROCESSING

ANALOG–DIGITAL CONVERTER

REGISTER

DIGITIZED PICTURE INFORMATION

STORING THE PICTURES

Images are stored on removable media like diskettes, flash memory cards or PC or PCMCIA cards for transferring to a computer. Under the cover of bigger-capacity cards is actually a hard disk with a circuit board for disk control. The more pixels in an image, the more storage space it takes up. The 2-million-pixel images produced by this camera each take up about 6 megabytes of space.

ELECTRONS VS. CHEMICALS

CAPTURING DIGITAL IMAGES
ANALOG-DIGITAL CONVERSION

C.C.D.

LENS

In digital photography the image, inverted and focused by the lens, is cast on a light-sensitive charge-coupled device.

FILM-BASED PHOTOGRAPHY
DEVELOPING AND PROCESSING

FILM

LENS

Layers of emulsion are sensitive to blue, green and red light. Chemical changes caused by the exposure to light become visible when film is developed.

Source: Kodak

Mika Gröndahl

Digital Cameras Harness Chip Technology to Do the Job of Film

By Jed Stevenson

Once considered an engineering folly in a world of film and video cameras, digital still cameras have become a booming industry.

Kodak introduced its first digital camera in 1991, an expensive model bought mostly by photojournalists — and well-heeled ones at that, because it cost more than $20,000. Four years later, Kodak offered the first digital camera it marketed to less affluent consumers: the DC40, which cost less than $800. Now Kodak, just one of many camera companies going digital, sells a complete digital line. The industry is growing fast.

At the heart of every digital camera is an image-sensing chip. It performs the same function as film — creating an image when light coming through the lens falls on it. But unlike film, which undergoes a chemical change when exposed and thus can only be used once, an image-sensing chip simply creates an electrical signal that can be converted to digital form, processed, stored, sent to a computer and manipulated. Best of all, the chip can keep making these signals indefinitely.

Digital cameras tend to be more expensive than film cameras because of the complexity of manufacturing the chips. And chips that can produce bigger and sharper photographs — that have more picture elements, or pixels, for a given area — cost more than lower-resolution chips.

As chip makers have managed to cram more pixels on chips, resolution has improved and prices have fallen. A basic consumer "point and shoot" model with a resolution up to about a megapixel (a million pixels) can now be had for a couple of hundred dollars. This resolution is fine for standard-size prints and is more than adequate for most computer applications, like sending family snapshots on the Internet and illustrating Web pages. For more money, consumers can buy cameras that produce 3- or 4-megapixel images — resolutions that were practically unavailable, at any price, in the early days of digital imaging.

Most digital cameras have used a type of image-sensing chip called a charge-coupled device, or C.C.D. While these chips can deliver enough pixel density to make a good photograph look good, they have one inherent problem. They are stand-alone chips, meaning they have to be wired into a circuit board on the camera that contains the processor and other chips for treating the digital signal. That makes manufacturing more complicated and expensive, and means that a C.C.D. camera consumes more power.

A different kind of imaging chip, called CMOS, for complementary metal-oxide semiconductor, doesn't have those problems. CMOS chips are now showing up in some cameras.

With CMOS designs, the image sensor, processor, filters and other electronic components can all be part of one chip. Because CMOS chips are manufactured using the same methods that are used to make hundreds of millions of conventional processing and memory chips, fabrication costs are low for the chips themselves. And the "camera on a chip" approach can reduce the cost of manufacturing the camera and significantly cut power consumption.

CMOS chips have been around for about 30 years, but they have needed to be extremely large to deliver the same pixel density as a C.C.D. chip. Improvements in pixel density have reduced the size and cost of these chips, and it would appear to be only a matter of time before they replace C.C.D.'s in many cameras.

Marc J. Loinaz, a researcher at Bell Labs, part of Lucent Technologies, said that as CMOS chips became smaller, digital cameras would eventually be limited in size only by lens sizes. "If there ever is a Dick Tracy video watch," he said, "it'll use CMOS chips."

Hold Still, but Not for Long

Most ear thermometers sold to consumers use a pyroelectric infrared sensor, which uses a shutter to take a snapshot-like sample of the infrared radiation coming from the eardrum. The signal from the detector is amplified and sent to a microprocessor in the barrel of the thermometer, which calculates the body temperature.

1
INFRARED RADIATION
Infrared radiation is emitted by the eardrum and bounces off the mirrorlike walls of an ear thermometer's probe.

2
SHUTTER
When the thermometer's Start button is pressed, a shutter inside the probe slides open, allowing infrared radiation to pass from the eardrum to the surface of the pyroelectric sensor for a brief time.

3
SENSOR
A pyroelectric infrared sensor is a flat ceramic plate with electrodes on either side. When the infrared radiation strikes the sensor, it warms the surface of the sensor by a few thousandths of a degree, changing its electrical conductivity. That produces an electrical signal proportional to the magnitude of the infrared signal.

INFRARED RADIATION ENTERS

SHUTTER OPENS

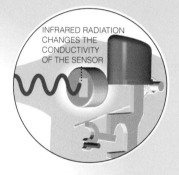

INFRARED RADIATION CHANGES THE CONDUCTIVITY OF THE SENSOR

BY EAR
The ear is a good site for accurate body temperature measurements because it is close to the hypothalamus, an area at the bottom of the brain that controls body temperature. The ear is also not cooled or warmed by eating, drinking or breathing. It is important to aim the probe directly at the eardrum to get a correct temperature reading.

START BUTTON

PROBE

can plus

ELECTRONICS

INFRARED RADIATION

4
ELECTRONICS
After the electrical signal is amplified, it travels to an electronic circuit in the barrel of the thermometer. A device there sends the analog signal to an analog-to-digital converter. The digital signal can then be read by the thermometer's microprocessor, which calculates the body temperature (taking into account things like the temperature of the probe itself).

5
DISPLAY
The microprocessor sends its result to a liquid-crystal display screen at the front of the thermometer barrel.

Source: Braun

Mika Gröndahl; Photograph by Naum Kazhdan

Smart Thermometers Make Life Easier for the Sick and Squirmy

By Catherine Greenman

In the late 1800's, if a patient was thought to have a fever, a nurse or doctor might well have wedged a mercury thermometer a foot and a half long into the patient's armpit and waited. And waited.

When doctors started using mercury thermometers in the mid-19th century, the instruments were so large that the mercury typically took almost a half-hour to heat up for a temperature reading. That is unthinkable for people today, who have become accustomed to getting a temperature reading in just seconds with an instant ear thermometer.

Manufacturers of ear thermometers say the ear is the best place to measure body temperature because the eardrum is a couple of inches from the hypothalamus, the area of the brain that, among its many duties, regulates body temperature.

Ear thermometers, which never actually touch the eardrum, are slowly replacing mercury and other electronic thermometers that must be in contact with the body.

But how does an ear thermometer achieve an accurate reading? And how does it do so without using a drop of mercury?

When something is warm, it gives off energy. The hotter an object, the higher the frequency of the electromagnetic radiation it gives off. Very hot objects, like light bulbs, radiate what people call light — electromagnetic radiation in the visible portion of the spectrum, from violet to red, with wavelengths of about 350 to 800 nanometers (a nanometer is a billionth of a meter).

The hotter a light bulb filament gets, the greater the intensity of the light. Reducing the heat changes the color of the light. Dimming a light bulb, for example, changes the color of the light from yellowish to reddish.

Just like the filament of a light bulb, the human body radiates heat, which can be detected as infrared radiation, or infrared light.

Although infrared light cannot be seen by the human eye, it can be detected by sensors designed to measure the radiation frequencies outside the visible range. So it is possible to determine someone's temperature by measuring the amount of infrared radiation being emitted.

Ear thermometers operate on this principle, calculating temperatures by measuring the infrared radiation that emanates from the eardrum and its surrounding tissue. The most widely sold ear thermometers for home use contain something called a pyroelectric sensor, a detection device that responds to infrared radiation.

The first pyroelectric ear thermometer, called the Thermoscan, was invented by Jacob Fraden, a biomedical electronic engineer in San Diego, and introduced to hospitals and doctors' offices in 1990. Now they are in wide use by professionals and, despite their relatively high cost, in the homes of happy parents who never want to pick up another fragile tube containing mercury.

A pyroelectric ear thermometer works like this: When you insert the probe into the ear canal and press the trigger button, a shutter in the probe (like a camera shutter) snaps open. Infrared radiation from the eardrum travels down the barrel of the probe and hits the pyroelectric sensor. The absorption of the infrared energy changes the sensor's electrical conductivity.

The sensor then sends an electrical signal indicating the amount of heat to a circuit board, where it is amplified, converted into a digital signal and sent to a microprocessor, which ultimately turns it into a digital temperature readout.

Another kind of sensor used in ear thermometers, called a thermopile, performs a continuous measurement of emitted infrared energy, rather than taking a snapshot. Because a thermopile sensor takes an average of five temperature readings, it must be left in the ear several seconds longer than a pyroelectric ear thermometer.

Regardless of what type of sensor is used to determine body temperature, every ear thermometer contains either a four-bit or eight-bit microprocessor. Using a complex mathematical approach that takes into account the ambient temperature, the microprocessor converts the signal measuring the infrared radiation into degrees Fahrenheit or Celsius, and displays the number on a screen.

The new thermometers have made life easier for sick children and are now helping sick animals, too. After Dr. Fraden sold Thermoscan to the Gillette Company in 1995, he developed a veterinary version, now marketed by the Advanced Monitors Corporation. The thermometer's probe is longer because horses, cows and dogs, for example, have deeper ear canals than humans (as well as body temperatures that average several degrees higher than humans).

What's That You Say?

The sounds people hear range from low frequencies (a bass guitar) through the midrange (the human voice) and up to high frequencies (chalk on a blackboard). Perfect hearing would be uniformly sensitive to sound at frequencies from 20 to 20,000 hertz (vibrations per second), but most people do not have perfect hearing.

LOW SOUNDS   HIGH SOUNDS

VOWEL SOUNDS   *(SPEECH)*   CONSONANT SOUNDS

250   500   1,000   2,000   4,000   8,000   16,000

FREQUENCY (hertz)

SHARON WOULD LIKE TO MEET YOU FOR DRINKS AT FIVE AT SIXTY-SIXTH AND PARK.

As people get older, they tend to lose sensitivity to higher frequencies. That means consonants — especially soft initial and final consonants, like "s" and "th" — are harder to hear because they are spoken at higher frequencies than vowels. Hearing aids can be adjusted or programmed to make the consonants easier to hear.

AUDIOGRAM: A PICTURE OF HEARING LOSS ▶

This audiogram shows the hearing of a person who cannot hear well in the mid- and high-frequency ranges. To compensate, a hearing aid would moderately augment incoming midrange sounds and strongly increase high-frequency sounds.

Source: Widex

LOW SOUNDS   HIGH SOUNDS

250   500   1,000   2,000   4,000   8,000   FREQUENCY (hertz)

MID-RANGE BOOST

HIGH-RANGE BOOST

NORMAL HEARING

VOLUME INCREASED BY HEARING AID

VOLUME PERCEIVED IN HEARING TEST INDICATES HEARING PROBLEMS

HEARING-LEVEL VOLUME (in decibels)

Hearing Aids Put Digital Audio Processing in the Ear

By Howard Alexander

The earliest hearing aid was simplicity itself and is still in use: a cupped hand behind the ear. But technology has made hearing aids much less obtrusive and much more helpful. While early hearing aids tried to improve upon the external ear, with devices to direct the sound coming from the front into the ear of the listener and block noise from the sides and rear, modern devices fit surreptitiously behind or into an ear while augmenting the sound coming in.

In the 17th century, artisans made metal "ears" to fit over the natural ears. Ear trumpets, cones that could be aimed at the source of a sound, came into use in the early 1800's. By the end of that century, the acoustic horn became available. It was a flexible tube with two ends — one cone shaped to catch the sound, the other gently tapered to fit inside the ear.

The approach changed from art to engineering in this century. In 1913, Siemens became one of the first companies to offer electronically amplified hearing aids. These early units were large and not very portable — about the size of a tall cigar box — but they had a speaker that fit into the ear. Acousticon's Model 56 in the mid-1920's was one of the first really portable units, although it was quite heavy. Some of the technological advances of World War II showed up after the war in hearing aids. Zenith's pocket-sized Miniature 75 helped set a trend toward smaller and more effective prosthetics.

Transistors led to startling improvements in hearing aids. Tiny transistors in newly developed miniature microphones could be built into the frames of eyeglasses along with better, smaller batteries. Mounting a hearing aid behind the ear, as in Zenith's Diplomat of the late 1950's, is called B.T.E. in the industry. The next step was to put hearing aids in the ear (I.T.E.), then, as they got even smaller, partly into the ear canal (I.T.C.). Now hearing aids can fit completely into the ear canal (C.I.C.).

SHARON WOULD LIKE TO MEET YOU FOR DRINKS AT FIVE AT SIXTY-SIXTH AND PARK.

How hearing works

THE EAR
The outer ear, or pinna, helps protect the ear and directs sound into the ear canal. Sound travels down the ear canal and hits the eardrum, causing it to vibrate. The motion is transferred through three small bones called the hammer (malleus), anvil (incus) and stirrup (stapes) to the inner ear. Attached muscles tend to damp excessive motion from loud sounds to protect the inner ear. The vibration is transmitted from the stirrup to the fluid within the cochlea.

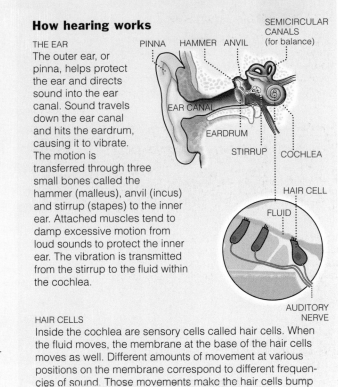

HAIR CELLS
Inside the cochlea are sensory cells called hair cells. When the fluid moves, the membrane at the base of the hair cells moves as well. Different amounts of movement at various positions on the membrane correspond to different frequencies of sound. Those movements make the hair cells bump into a membrane above them and bend, triggering signals that eventually reach the brain through the auditory nerve and are interpreted as sound.

Mika Gröndahl; Photograph of hearing aid by Widex, others by Igor Vishnyakov

DIGITAL HEARING AID THAT FITS COMPLETELY IN THE EAR CANAL (C.I.C.)

COMPONENTS:

MICROPHONE
The microphone collects the sound waves entering the ear canal.

PROCESSING CHIP SPEAKER BATTERY

A microprocessor takes into account the wearer's specific hearing problems and handles the processing of the incoming sound accordingly. It makes it possible to adjust settings to match particular environments and incoming sound characteristics. It also avoids feedback, the squealing sound that occurs when the microphone picks up some of the sound from the hearing aid's speaker.

SPEAKER
The speaker, or transducer, transforms the digital signal from the microprocessor into analog sound waves that can be heard.

Until the early 1990's, most hearing aids were analog and consisted of a microphone, an amplifier and a speaker (transducer). But with the availability of miniature computer components, as well as the audio-processing technology found in CD and DVD players, the digital hearing aid was born. Most hearing aids now have some digital components, and all-digital hearing aids are becoming popular.

Using microprocessors, digital hearing aids are able to manipulate the incoming sound to amplify the specific frequencies that a user is having trouble hearing. This feature, called equalization control, was rarely available with analog devices. One example of this kind of technology is the Senso, a fully digital unit that fits completely in the ear and is made by the Danish company Widex. A number of other companies also make digital hearing aids, including Siemens, a German company, and Telex, based in the United States.

Settings on digital hearing aids can be adjusted for specific environments. Going to a chamber music concert? Select setting No. 1. Meeting the boss for a business lunch in a noisy restaurant? Select setting No. 4.

What these settings do is adjust the equalization, volume and signal-processing functions. If you cannot hear high frequencies well but love chamber music, the appropriate setting would emphasize the high frequencies (for the violins) and give you a moderate increase in volume.

Lunch with the boss would require an emphasis on the middle range (for the human voice), a decrease in the high frequencies and some form of extraneous noise reduction.

Several companies are beginning to offer products with greater processing power that they say will provide a better match for a user's pattern of hearing loss. And cochlear implants are beginning to be used. These implants bypass some of the faulty nerve circuitry in the cochlea, a spiral-shaped structure in the inner ear, and allows some people who are deaf or who have severe hearing loss to have a degree of hearing. A number of companies, like the Cochlear Corporation, have developed cochlear implants, which are surgically connected to the cochlea. Researchers are also studying the possibility of brain implants or nerve regeneration as future remedies for hearing loss.

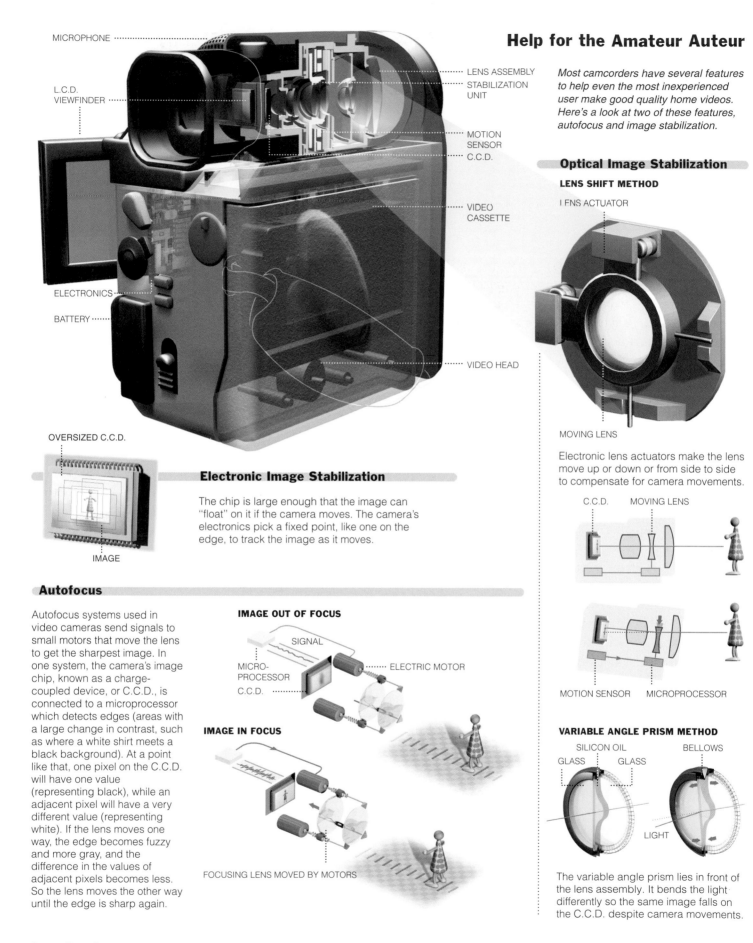

MICROPHONE

L.C.D. VIEWFINDER

LENS ASSEMBLY

STABILIZATION UNIT

MOTION SENSOR

C.C.D.

VIDEO CASSETTE

ELECTRONICS

BATTERY

VIDEO HEAD

OVERSIZED C.C.D.

IMAGE

Help for the Amateur Auteur

Most camcorders have several features to help even the most inexperienced user make good quality home videos. Here's a look at two of these features, autofocus and image stabilization.

Optical Image Stabilization

LENS SHIFT METHOD

LENS ACTUATOR

MOVING LENS

Electronic lens actuators make the lens move up or down or from side to side to compensate for camera movements.

C.C.D.   MOVING LENS

MOTION SENSOR   MICROPROCESSOR

Electronic Image Stabilization

The chip is large enough that the image can "float" on it if the camera moves. The camera's electronics pick a fixed point, like one on the edge, to track the image as it moves.

Autofocus

Autofocus systems used in video cameras send signals to small motors that move the lens to get the sharpest image. In one system, the camera's image chip, known as a charge-coupled device, or C.C.D., is connected to a microprocessor which detects edges (areas with a large change in contrast, such as where a white shirt meets a black background). At a point like that, one pixel on the C.C.D. will have one value (representing black), while an adjacent pixel will have a very different value (representing white). If the lens moves one way, the edge becomes fuzzy and more gray, and the difference in the values of adjacent pixels becomes less. So the lens moves the other way until the edge is sharp again.

IMAGE OUT OF FOCUS

SIGNAL

MICRO-PROCESSOR

C.C.D.

ELECTRIC MOTOR

IMAGE IN FOCUS

FOCUSING LENS MOVED BY MOTORS

VARIABLE ANGLE PRISM METHOD

SILICON OIL

GLASS   GLASS

BELLOWS

LIGHT

The variable angle prism lies in front of the lens assembly. It bends the light differently so the same image falls on the C.C.D. despite camera movements.

Sources: Canon, Sony

Mika Gröndahl

Camcorders Allow a Steady Shot, Even Without a Steady Hand

By Roy Furchgott

Although jittery, frenetic images were the height of film artistry in the early 90's, it isn't a good technique for immortalizing your 10-year-old's first soccer goal on videotape. And therein lies a problem. As camcorders have become smaller and lighter, with greater zoom capabilities, they have magnified the operator's tiniest tremors. Even breathing can result in a jerky image.

But the camcorder industry has responded with technology that can reduce the jiggle. Digital equipment can read a recorded image and make adjustments to compensate for movement, while mechanical optical systems can keep an image centered and in focus by using motion sensors that move the lens to counter unwanted motion.

The problem isn't unique to the camcorder industry. Hollywood filmmakers have long struggled to move cameras without introducing unwanted quivers into the picture. Movie makers often put a camera on a heavy dolly that rolls on smooth railroad-like tracks. "This has always been a difficulty," said Robert Dickson, a historian with the American Film Institute. "The camera has to be pushed carefully."

Cinematographers began to experience the problems that would face home video enthusiasts when lighter cameras, about 35 pounds, began to replace their 80-to-110-pound predecessors in the early 1970's. John Hora, a cinematographer, said he confronted the problem with dolly shots on "The Twilight Zone," a movie made in 1983.

"You could see a little motion if you look right at the edge of the frame," Mr. Hora said. In later movies, he went back to heavier equipment. "If you run over a little piece of tape on the track with a heavy piece of equipment, you won't feel it," he said. "It's like being in a Cadillac and hitting a bump — you float right over it. A lighter unit is like a sports car — you'd feel the bump." And see it as well.

The popularity of the handheld shot created more trouble for camera operators, who had to develop a gliding Groucho-style walk to keep from shaking the camera. "You have to walk with your knees bent," Mr. Hora said. "There's a real skill to it."

That changed with the Steadicam, a system that suspends a camera on spring-loaded arms that are attached to a vest worn by a camera operator. But even the amateur version of the Steadicam costs more than $1,300.

In 1992, the home camcorder industry recognized the shaky-picture problem: cameras had become so light that hand tremors would move them and longer lenses made the problem worse. A small movement like clearing your throat can make it seem like an earthquake struck the scene when you are recording on videotape at a magnification of 22X.

A digital solution, developed about 1996, lets the lens illuminate just part of the camcorder's imaging chip. When the camera moves, the whole frame changes position on the chip, but it still registers in its entirety. The camera's computer picks a fixed point in the picture, like the outside edge of the frame, then stabilizes that edge as a fixed point on the recording.

But the system is fallible. "You are not sensing motion on the part of the camera," said Mark Weir, product planning manager for digital imaging at Sony Electronics. "You are seeing how the image is changing and assuming its motion." If the camera is perfectly still and a hand is waved past it quickly, it might pan to the side because it assumes that the camera has moved.

The mechanical solution uses a complex moving prism to keep the image centered on the chip. A mechanism typical of camcorders is Canon's variable-angle prism, actually a pair of prisms that move in relation to each other. The movement of the prisms is controlled by actuators that get signals from a microcomputer that monitors motion, direction and intensity.

This system adjusts for any shaking of the camcorder, smoothing out even distant shots. It tends to produce a more accurate, clearer picture than digital correction, but it's also more costly.

Remote Controls from Yesterday . . .

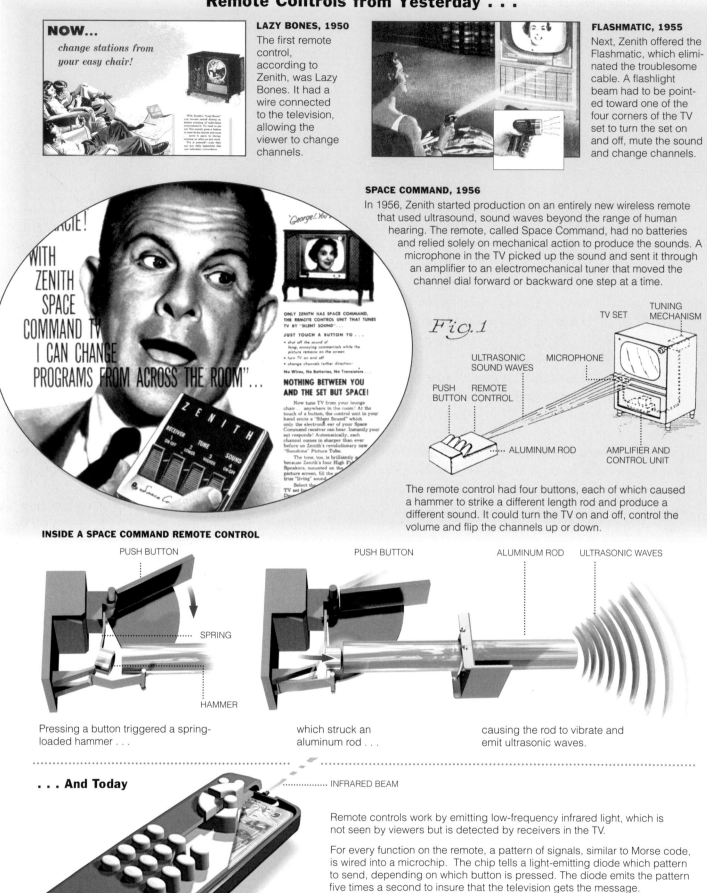

NOW...
change stations from your easy chair!

LAZY BONES, 1950
The first remote control, according to Zenith, was Lazy Bones. It had a wire connected to the television, allowing the viewer to change channels.

FLASHMATIC, 1955
Next, Zenith offered the Flashmatic, which eliminated the troublesome cable. A flashlight beam had to be pointed toward one of the four corners of the TV set to turn the set on and off, mute the sound and change channels.

SPACE COMMAND, 1956
In 1956, Zenith started production on an entirely new wireless remote that used ultrasound, sound waves beyond the range of human hearing. The remote, called Space Command, had no batteries and relied solely on mechanical action to produce the sounds. A microphone in the TV picked up the sound and sent it through an amplifier to an electromechanical tuner that moved the channel dial forward or backward one step at a time.

Fig.1

TV SET
TUNING MECHANISM
ULTRASONIC SOUND WAVES
MICROPHONE
PUSH BUTTON
REMOTE CONTROL
ALUMINUM ROD
AMPLIFIER AND CONTROL UNIT

The remote control had four buttons, each of which caused a hammer to strike a different length rod and produce a different sound. It could turn the TV on and off, control the volume and flip the channels up or down.

INSIDE A SPACE COMMAND REMOTE CONTROL

PUSH BUTTON
SPRING
HAMMER

PUSH BUTTON
ALUMINUM ROD
ULTRASONIC WAVES

Pressing a button triggered a spring-loaded hammer . . .

which struck an aluminum rod . . .

causing the rod to vibrate and emit ultrasonic waves.

. . . And Today

INFRARED BEAM

Remote controls work by emitting low-frequency infrared light, which is not seen by viewers but is detected by receivers in the TV.

For every function on the remote, a pattern of signals, similar to Morse code, is wired into a microchip. The chip tells a light-emitting diode which pattern to send, depending on which button is pressed. The diode emits the pattern five times a second to insure that the television gets the message.

Source: Zenith Electronics Corporation

Mika Gröndahl

The Gadget That First Taught a Nation to Surf

By Lisa Napoli

Back when the culture of instant digital gratification was just a gleam in a futurist's eye, a revolutionary device was invented that was one of the first steps toward technological power for consumers: the remote control.

It changed everything. "The twitch of our fingers on TV remotes was the first sign of slumbering interactive desires," said Paul Saffo, a director at the Institute for the Future, in Menlo Park, Calif. "The remote control wasn't just a new way to change channels. It changed TV as a medium. Suddenly you didn't have to watch shows."

Viewers could channel surf instead. They could watch several programs at once, have their senses assaulted by sounds and images and, best of all, they could skip the commercials.

The impetus for the remote control was as much to silence advertisements as to let viewers remain on the couch. The founder of the Zenith Electronics Corporation, Eugene F. McDonald Jr., said he thought advertisements would be the death of the nascent television medium.

"He wanted to prove his point by making commercials unprofitable," said Dr. Robert Adler, the Zenith engineer who created the first practical wireless remote control, which used ultrasonic signals and went on the market in 1956. More than nine million televisions with ultrasonic remote controls were sold in 25 years, according to Zenith.

But the remote control had a different effect. It forced advertisers to create commercials that were more clever so they could capture the attention of viewers as they clicked through the channels.

"Before the remote, the advertisers had you as their hostage," Mr. Saffo said. "They knew you weren't going to get up."

Dr. Adler's creation, called the Space Command, was mechanical. It literally clicked, or pinged, when a button was pushed. It used an ultrasonic beam, outside the range of human hearing, to turn a set on or off or change the channel.

The price for a set and a remote control ranged from $259.95 for a tabletop model to $550 for a high-end console — hefty price tags back then. Advertisements for the Space Command proclaimed that there was "nothing between you and the set but space!" Some ads went on to say: "Is it magic? It's like nothing you have ever seen before — anywhere!"

It wasn't magic, but it was a big step beyond Zenith's earlier remote controls. The first such control, marketed in 1950, was the inelegant but aptly named Lazy Bones. It was tethered to the set by a long wire, which made it a living-room hazard. The next version, in 1955, was wireless. It was essentially a flashlight that changed the channels when it was focused on sensors built into the cabinet of a set. Thus its name: the Flashmatic.

"On a sunny day you had to turn it off because the tuner would just go crazy," Dr. Adler recalled in a telephone interview. Disadvantages aside, more than 30,000 Flashmatics were sold.

The first remote controls seemed to be ahead of their time because there were few stations to flip through in the 50's. "It's surprising how early this technology was created," said Ronald C. Simon, a curator at The Museum of Television & Radio in Manhattan.

It took awhile for the remote control to catch on. Among the televisions sold in 1976, 9.5 percent had remote controls; by 1990, 90 percent of them did. But the advent of the remote control was a big step toward the couch potato culture to come. Viewers could stay put while they watched, orchestrating their viewing like maestros with batons, while what they watched came at them at a faster pace. Over time, there were more channels and shows than ever before, more than it seemed possible to navigate.

"There's a huge magnetic force in the culture and it goes toward speed, and the remote control was an essential element in that direction," said Todd Gitlin, a professor of media studies at New York University. " 'Move it along' — that's what matters. The speed of movement and the juxtaposition of different images is absolutely central to how people are living now."

Taking over the TV

The earliest form of parental control over children's viewing habits resided in a simple piece of technology: the on-off switch. But parents who want to exercise control with more finesse can use a television equipped with a V-chip to block only programs with certain ratings. Broadcasters can transmit ratings along with their programs by using a part of the television signal that used to be blank.

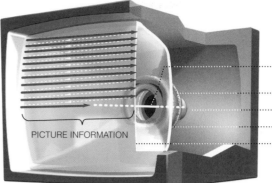

ELECTRON BEAM

ELECTROMAGNETS STEER THE ELECTRON BEAM

ELECTRON GUNS

VACUUM TUBE

PHOSPHOR COATING

PICTURE INFORMATION

A SIGNAL WITHIN A SIGNAL

Television pictures are drawn line by line by three electron beams inside the television set's cathode ray tube.

The three beams, one each for red, green and blue colors, start from the top left of the screen, as seen by the viewer, and scan back and forth to draw the lines that create the still image (or frame).

Frames are drawn 30 times a second. The guns that fire these beams turn off when they reach the bottom right of the screen so that they can get to the top left again without drawing a line across the screen. To allow time for the guns to be repositioned, broadcasters transmit a pause called the vertical blanking interval, or V.B.I., in part of the 525-line frame.

V.B.I.

EXTRA CIRCUITRY

The V-chip, a special component added to any television with a screen larger than 13 inches, makes use of data contained in the V.B.I.

Mika Gröndahl

The V-chip: Parental Guidance for the Television

By Matt Lake

Everyone has opinions about television sex and violence, and opinions about the technology to block it, the V-chip, run equally strong. To its supporters, the V-chip is a triumph of technology that allows parents to keep material that runs contrary to their tastes or beliefs from being viewed by their children. To critics, it is a tool for government interference that falls only slightly short of censorship.

But to television set manufacturers, the V-chip is a bit of a nuisance. In 1999, the Federal Communications Commission required that half of all new sets with screens larger than 13 inches include technology that lets consumers block programs based on their content. By 2000, all sets that big were subject to the rule.

The V-chip had its beginnings in 1990, when Congress passed two acts aimed at regulating the television industry. The Television Violence Act declared that broadcasters would not violate antitrust laws if they worked together to reduce violence on television, and the Television Decoder Cir-

cuitry Act mandated that within three years, television receivers should be capable of decoding signals for closed-captioning. In those three years, an industry task force and the communications commission worked to establish specifications for transmitting and decoding data. The approaches went beyond closed-captioning and were based on a broadcasting feature called the vertical blanking interval.

Television pictures are transmitted as lines of information that are drawn across the screen by cathode-ray guns in the picture tube. The vertical blanking interval, part of every television picture, is a few empty lines broadcast at the end of every frame so the cathode-ray guns can return to the top of the screen to start drawing the next frame. Since the interval contains no picture information, it has bandwidth to spare.

Some of that bandwidth is set aside for closed-captioning, but that still leaves plenty of bandwidth. In the 1970's, the British Broadcasting Corporation and the Independent Broadcasting Association in Britain began using the extra

PLENTY OF ROOM

Since the V.B.I. contains no picture information, it has bandwidth to spare: up to 57,600 bits. Broadcasters use part of this bandwidth for closed-captioning and timing signals. In response to the Telecommunications Act of 1996, the consumer electronics industry decided to use part of the V.B.I. to transmit a show's rating. V-chips can interpret the V.B.I. signals, which include movie ratings (like PG and NC-17) and the TV Parental Guidelines ratings that flash on screen for the first 15 seconds of many TV shows.

PARENTS AS PROGRAMMERS

The V-chip does not block any shows until it is programmed to do so. Using a remote control and on-screen menus, viewers can program the V-chip to block any broadcast with a rating for material they find objectionable, like violence, suggestive dialogue or profanity. That changes the setting of a register address in the chip. The change is stored in reprogrammable nonvolatile memory, which means that the setting is retained even when the power is turned off but can be altered at another time.

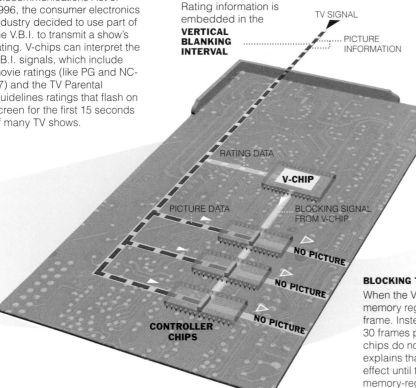

Rating information is embedded in the **VERTICAL BLANKING INTERVAL**

TV SIGNAL
PICTURE INFORMATION
RATING DATA
V-CHIP
PICTURE DATA
BLOCKING SIGNAL FROM V-CHIP
NO PICTURE
NO PICTURE
NO PICTURE
CONTROLLER CHIPS

THIS CHANNEL IS NOT APPROVED FOR VIEWING

BLOCKING THE SIGNAL

When the V-chip detects a broadcast rating that matches a setting in its memory registers, it instructs the controller chips not to display the frame. Instead of having the electron guns in the cathode ray tube draw 30 frames per second to display the program, the picture controller chips do not allow the program to be shown. Instead, a message explains that the broadcast is being blocked. The blocking stays in effect until the setting in the broadcast's V.B.I. does not match a memory-register setting that calls for a program to be blocked.

---

bandwidth to supply interactive text-based news and weather services, called Ceefax and Oracle.

In the United States, part of the blanking interval was set aside to carry program identification information, schedules and emergency broadcast functions. In 1993, the Television Violence Reduction Through Parental Empowerment Act was passed, and the term V-chip was introduced. To comply with the act, broadcasters could opt to transmit ratings for all programs (except sports and news) on the blanking interval, and V-chips would be included in new television sets and set-top boxes to interpret the ratings.

A coalition of the Motion Picture Association of America and cable and network broadcasters came up with a set of ratings in January 1997. The system has six categories: TV-Y, TV-Y7, TV-G, TV-PG, TV-14 and TV-MA. The two Y ratings designate programs for young children; TV-G is for general audiences, and TV-PG calls for parental guidance for children. TV-14 indicates shows for those 14 and older, and TV-MA is for mature audiences.

By programming a V-chip to reject shows with a particular rating, parents block all programs in that group. But the broad ratings have subdivisions that indicate whether a show includes violence (graphic violence or cartoonlike fantasy violence), sexual content, profanity or suggestive dialogue. It is possible to block programs rated TV-14 because of violence but allow those given the rating for profanity. The V-chip can also block broadcast movies that carry a Motion Picture Association of America rating — G through NC-17 — although the movie ratings are not subdivided to indicate whether a show has profanity or sexual content.

The system is clever, but it is not perfect. The chips are hard-wired to recognize signals in specific places in the blanking interval. If the ratings system changes substantially, or if broadcasters choose not to adopt the voluntary ratings scheme, violent content can still make it past the V-chip. And since sports and news broadcasts are exempt from ratings, parental control over viewing is limited, although most television sets that include V-chips also allow the blocking of whole channels. A parent who is so inclined can pick up the remote control and block CNN and ESPN.

Sound That Seems to Come From Out of Nowhere

Virtual surround sound systems use electronics to create an illusion that sounds are coming from many sources rather than just the two speakers at the front of a room. Before the electronic devices can be developed, however, audio engineers must first measure the effects that the body — in particular, the ears — have on sound.

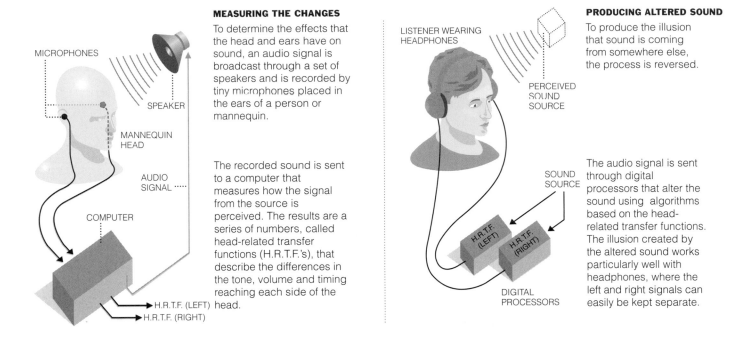

MEASURING THE CHANGES

To determine the effects that the head and ears have on sound, an audio signal is broadcast through a set of speakers and is recorded by tiny microphones placed in the ears of a person or mannequin.

The recorded sound is sent to a computer that measures how the signal from the source is perceived. The results are a series of numbers, called head-related transfer functions (H.R.T.F.'s), that describe the differences in the tone, volume and timing reaching each side of the head.

PRODUCING ALTERED SOUND

To produce the illusion that sound is coming from somewhere else, the process is reversed.

The audio signal is sent through digital processors that alter the sound using algorithms based on the head-related transfer functions. The illusion created by the altered sound works particularly well with headphones, where the left and right signals can easily be kept separate.

Labels in diagram: MICROPHONES, SPEAKER, MANNEQUIN HEAD, AUDIO SIGNAL, COMPUTER, H.R.T.F. (LEFT), H.R.T.F. (RIGHT), LISTENER WEARING HEADPHONES, PERCEIVED SOUND SOURCE, SOUND SOURCE, H.R.T.F. (LEFT), H.R.T.F. (RIGHT), DIGITAL PROCESSORS

The Sweet Deception of Virtual Surround Sound

By Eric A. Taub

The attempt to turn a room of your house into a home theater rivaling the local megaplex demands a lot of equipment: DVD players, giant televisions and audio receivers that double as electronic control centers.

Also required are speakers — lots of them. When it comes to reproducing natural sound, the most realistic effect would come from having an infinite number of speakers, which would mimic the transmission of audio waves from all points in the environment. As a compromise, the Dolby Digital sound standard, which is now used for DVD players, laser discs and digital and satellite television, calls for six speakers: left and right front, center front (for dialogue) and left and right rear, as well as one for low-frequency effects, which can be placed just about anywhere in the room. Such a setup can make the action in an action thriller sound as if it is happening two inches away.

But many households lack the space or money for even six speakers. The solution is to use virtual surround sound, a technology that fools the brain into thinking that there are speakers to the sides and the rear when you actually have as few as two in front of you.

Virtual surround sound is incorporated into most DVD players and audio receivers and into some digital televisions. There is also Dolby Headphone technology, from Dolby Laboratories, which can turn ordinary stereo headphones into a five-channel listening experience.

In a room with proper acoustics, virtual speakers (also called phantom speakers) can sound like the real thing. "Sixty percent of our test subjects prefer hearing surround sound through virtual speakers than through real ones," said Alan Kraemer, engineering director for SRS Labs, developers of the TruSurround virtual surround system used in many DVD players.

Scientists have long known how to make sounds appear to be emanating from somewhere other than their actual source. The brain perceives the direction of a sound not from its location but from several other criteria, like the time it takes for sound perceived by one ear to be heard by the other one and the balance of the various frequencies that actually reach each ear.

For example, the structure of the ear alters some high-frequency components (known as overtones) of sounds pro-

PUTTING IT ALL TOGETHER: A HOME THEATER

A virtual surround sound system combined with a large-screen television can create the illusion of being in a theater. Although the system uses only two speakers, when the listener is sitting in the "sweet spot" (directly in front of the speakers) the impression will be that the sound is coming from all directions. Systems try to eliminate the problem of cross talk — the signal intended for one ear also reaches the opposite ear — by using a digital filter that broadcasts a signal that cancels out, for example, the left-ear signal when it reaches the right ear (as shown at far right).

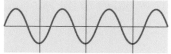

SYSTEM SPEAKERS

■ Real ◇ Phantom

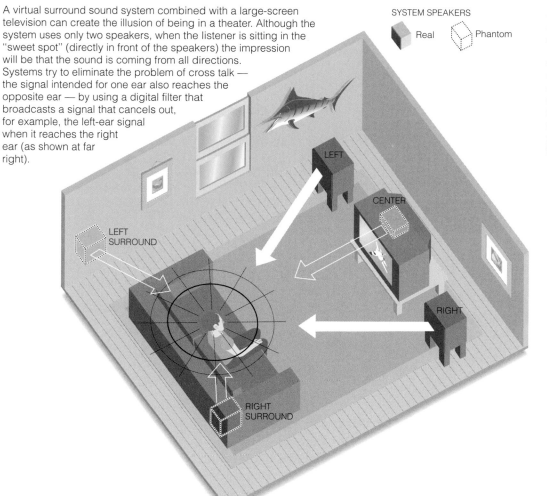

LEFT SURROUND

LEFT

CENTER

RIGHT

RIGHT SURROUND

CANCELING CROSS TALK

Sound can be canceled by taking a signal and delaying it so there are two signals that are mirror images of each other — when one signal is at its peak, the other is at its trough. The two signals cancel each other out. A simplified version of this is shown below; in reality, the signals that the ear receives are very complex, requiring precise filtering for proper cross-talk cancellation.

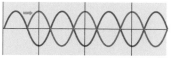

When a sound wave ...

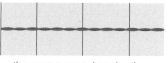

... meets its opposite...

... the waves cancel each other.

John Papasian

---

jected from the rear. The brain interprets the overtones not as different sounds but as coming from a different point.

Scientists determined the quality that each tone should have by placing tiny microphones in the ear canals of test subjects. A range of frequencies was played from different locations as instruments measured the frequency balance that reached each ear.

While big ears may be a cause for teasing, they are actually better at localizing sounds than smaller ones. To account for differences in ear sizes and shapes, many subjects were tested, and the results were averaged. The data were used to create formulas for altering the frequency characteristics of a sound so it will be perceived as if it were coming from another direction.

To create virtual surround sound, the digital audio signal is run through microprocessors and other electronic devices that alter the frequencies and the time that it takes the sound to reach each ear, according to these formulas. The result is two-speaker sound that seems to come from more than two locations.

The technology works well with headphones, as long as the listener's eyes are closed and the head stays facing one direction. But the listener quickly loses the sense of sitting in front of a symphony orchestra as soon as the eyes open and pick up visual cues about the room. If the head is turned, the orchestra seems to turn with the listener.

Even with external loudspeakers, there is a "sweet spot," an area where the auditory illusion of surround sound exists; move out of it, and the frequencies that your ears expect to hear are altered. If the speakers are placed incorrectly, or there are too many reflecting surfaces, the technology will not work because each sound is received several times, each time with a different balance of frequencies.

If you have 10 people over for a night at the movies, don't expect to wow them with new phantom speakers. "Virtual surround sound is a solitary experience," said Floyd Toole, vice president for acoustical engineering at Harman International. "If you're looking to reproduce sound in a social situation, you can't use it because it's too localized."

Through Thick and Thin

Capacitive stud finders work by measuring density differences in different parts of a wall. They make use of the principle of capacitance, the ability of a material to store an electrical charge. That ability depends upon the material's dielectric constant — essentially, how good an insulator it is. The dielectric constant varies depending upon the type of material and its thickness. So when the stud finder moves from an area of just wallboard to an area of wallboard and stud, the dielectric constant, and thus the capacitance, changes.

LED

CHIP

SWITCH

BATTERY

CALIBRATION

When you place a stud finder against a wall and turn it on, it creates a weak electrical field that builds up a charge in the wall. A metal plate in the finder, along with a microprocessor and a timer, measures how long it takes to reach a given level of charge. This becomes the baseline reading for the wall's thickness.

FALSE ALARMS

Most capacitive stud finders work well with walls made of wallboard as thick as three-quarters of an inch, although more expensive models can detect studs through even thicker material. But the devices can run into trouble under several conditions:

■ They can give false readings when used on older walls made of plaster and wooden lath. That is because the back of a plaster wall is not even.

■ Newer plaster walls, which use metal mesh lath, can confound the detection circuitry.

■ Pipes and electric wiring can give false readings because they have different dielectric constants

OPERATION

1. As you move the stud finder across the wall, it continues to emit its electric field and measure the capacitance of the wall. As long as the device stays over material of the same density, the measurement stays about the same.

2. When the finder passes over denser material — a stud, for instance — it takes longer for the electric charge to build up to the target level. When this time difference becomes significant, the microprocessor triggers alert signals, usually a flashing light-emitting diode and a beep.

3. The stud finder detects a drop in capacitance when it reaches the end of the stud, triggering other alert signals. Some models have more than one metal plate measuring capacitance. By comparing readings from plates of different sizes, the point where density changes occur can be pinpointed precisely.

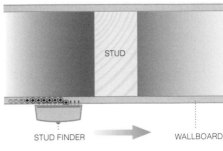

STUD

STUD FINDER   WALLBOARD

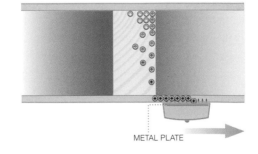

METAL PLATE

Source: Zircon Corporation

Mika Gröndahl

Stud Finders Take the Measure of a Wall

By Matt Lake

The next time you need to hang a picture on your wall or install some shelves, you might rely on aesthetics alone to decide the placement. Of course, if you do that, you risk being awakened in the middle of the night by a loud crashing sound as your object falls from the wall under its own weight.

That's because, as every contractor and do-it-yourselfer knows, you can't rely on wallboard or plaster alone to support something heavy. For real support, you need to fix it directly to a stud inside the wall. Contractors have tricks for finding the studs. Most involve rapping against the wall and listening for a change in the tone of the knock, then banging in a thin nail to verify the stud's location. Another old trick involves turning on an electric razor and dragging the butt end across a wall, noting the change in tone as you move from a wall cavity to a stud.

But in the 1970's, technology began to creep into the business of stud-finding. Devices appeared that eliminated the need for good hearing — and for a big tub of spackle to patch the wall when, inevitably, you ended up missing the stud anyway.

The earliest stud finders detected ferrous metals using compasslike pivoting magnets. Even though only commercial buildings use metal studs, magnetic stud finders, which are still available, work with wooden studs, too, by locating the metal nails used to mount the wallboard or wooden lath.

A newer type of stud finder works by detecting density changes in a wall. These capacitive finders can typically detect changes in wall density to a thickness of about three-quarters of an inch. Advanced models, like the StudSensor, manufactured by the Zircon Corporation of Campbell, Calif., have a setting that increases the sensitivity to a little more than an inch.

This works fine, at least in theory, for wallboard-covered walls, because the wallboard is thinner than the stud finder's depth range. But older plaster walls are often thicker and the thickness varies considerably, which can result in some false positive readings for the presence of studs.

A more serious drawback to the technology is that studs are not the only things in walls that register a different density. Pipes are much denser than plaster and lath (and studs), and wiring and conduit can also trigger a positive reading. Because it is especially ill-advised to knock nails into pipes and wires, some low-tech common sense must come into play when operating a stud finder. Any smart do-it-yourselfer or contractor has a mental checklist of items that confirm that a reading is really a stud.

The first rule of operation is to use the stud finder as its instruction book recommends, and that involves marking the edges of any stud reading with a pencil. As you continue across the length of the wall, you mark any stud readings as you go. You should be able to see a pattern of readings emerge — studs are usually 16 inches apart, and that is the pattern you should see. Any reading that seems to have a different width or an irregular placement is most likely not a stud at all, but some anomaly that should not have screws or nails driven into it.

On the horizon, there is a third technology being adapted to the problem of hidden studs: micropower impulse radar. This low-cost microwave radar technology was developed at the Lawrence Livermore National Laboratory in the 1990's. The laboratory and the electronics engineer who invented the technology, Thomas McEwan, obtained a patent on a radar stud detector in 1995, and the lab has since licensed the technology to several companies, some of which expect to market radar stud finders soon.

Micropower impulse radar has a considerably greater range than capacitive technology — between several inches and several feet deep. But finding studs is not the only planned use of the technology. It may also be used, among other things, for locating steel reinforcement in concrete pilings, as part of an in-vehicle device to warn drivers about vehicles in their blind spot, and to help find people inside collapsed buildings. That is all a far cry from banging picture hooks into walls, but it all boils down to the same technological problem of finding something in a space you can't see.

No Treasure Map Required

A portable metal detector uses an electromagnetic field to help treasure hunters search empty beaches and other spots for buried objects — a gold ring, perhaps, or, more likely, the occasional nickel or old pop-top.

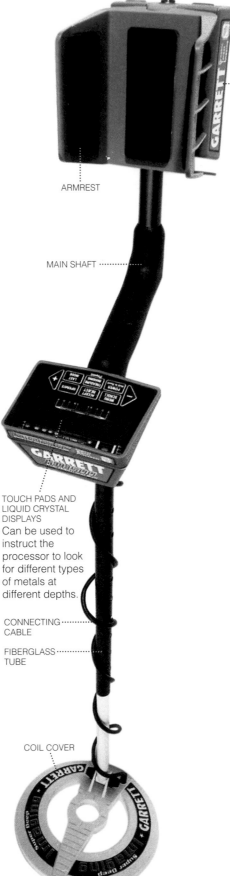

CONTROL BOX

ARMREST

MAIN SHAFT

TOUCH PADS AND LIQUID CRYSTAL DISPLAYS
Can be used to instruct the processor to look for different types of metals at different depths.

CONNECTING CABLE

FIBERGLASS TUBE

COIL COVER

TELLING DIFFERENT METALS APART

Some metal detectors can tell one metal from another. Highly conductive metals like copper and silver produce strong currents when they come in contact with an electromagnetic field. Others, like gold, lead and platinum, produce weaker currents. Some detectors can also discriminate between metal objects and highly magnetic ground minerals like iron, storing information about the signals that ground minerals produce for future reference.

Sources: Garrett Electronics; White's Electronics; Minelab Electronics; Dr. Richard C. Lanza, Massachusetts Institute of Technology

❶ CREATING THE ELECTROMAGNETIC FIELD

Electric current sent through a transmitter coil produces an electromagnetic field. In a Very Low Frequency, or V.L.F., metal detector, the field changes direction in cycles of 6 kilohertz (6,000 times per second) to 70 kilohertz.

The size of the transmitter coil may vary, affecting the size and shape of the electromagnetic field transmitted.

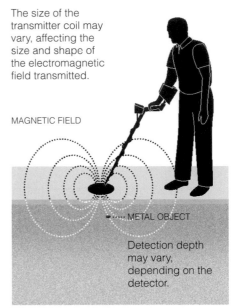

MAGNETIC FIELD

METAL OBJECT

Detection depth may vary, depending on the detector.

❷ WHEN METAL IS ENCOUNTERED UNDERGROUND

When the electromagnetic field created by the metal detector encounters a metal object, the field induces electric currents in the object, which in turn generates a smaller magnetic field. That secondary magnetic field, and the disturbance it causes, is what the metal detector detects.

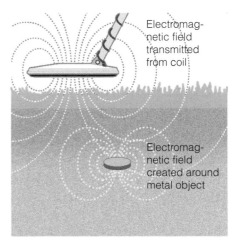

Electromagnetic field transmitted from coil

Electromagnetic field created around metal object

❸ ALERTING THE TREASURE HUNTER

The effect of the buried metal's own magnetic field generates an electric current in a second, detection coil. That current is amplified by the detector's electronics, and the control box emits a noise to alert the user. Some detectors have just one coil that serves both to transmit and detect.

Metal Detectors: Help for the Modern-Day Beachcomber

By Catherine Greenman

Every day in the summer, Ettore and Diana Nannetti start at 6 A.M. for Coney Island or another of New York City's famous beaches. They're trying to beat traffic, but not in order to nab a place in the sun. No, the Nannettis hit the beach to search for buried objects with metal detectors. When the monotonous tone in their headsets rises in pitch and volume, they know they've struck something. And by 10, when the sun is starting to climb in the sky and the tanning is getting good, they're gone.

"When I was young, I was very interested in history and archeology," Mr. Nannetti said. "This is almost an extension of a mini-archeological dig. You have no idea what you're going to find."

The Nannettis rarely find more than a few salt-caked coins or the occasional class ring, which they try to return. "We do it for the enjoyment, not for monetary value," said Mr. Nannetti, who counts an unopened six-pack of beer as one of his most memorable finds since he started using metal detectors in 1971.

In the hopeful world of metal detecting, it's an aluminum pull tab today, but maybe a gold ring tomorrow. The question for the average beachgoer who may occasionally see someone sweeping the sand is: How does a metal detector distinguish between the two?

All metal detectors are based on one principle: metal reacts to a magnetic field. A metal detector runs electric current through a transmitting coil to create an electromagnetic field that extends under the surface of the ground and changes direction several thousand times per second. If that field comes in contact with metal, it induces electric currents in the metal, which in turn create another magnetic field.

The effect of that secondary magnetic field is picked up by the metal detector's receiver coil. The detector's electronic components then signal the user with sound or a visual display. Metals that don't conduct electricity well but are easily magnetized can also generate a strong signal because they are magnetized by the detector's magnetic field.

Most hobbyists, known to their fellow searchers as detectorists or T.H.'ers (for treasure hunters), use newer Very Low Frequency detectors, meaning the magnetic field in the transmitting coil changes direction at a slower rate than in older models. Some V.L.F. models are able to reject highly magnetic minerals, like iron, that are simply part of the soil, which foiled the older machines. V.L.F. detectors can also take a guess at the depth of the object by the strength of the distortion signal. The larger the object, the stronger the current induced in the receiver coil.

Although many detector enthusiasts extol the thrill of the find and either return items or save them as collectibles, some sell what they find to dealers, collectors and pawnshops. "Found! 100 Gold Coins worth $36,000!" and "$4,375 Gold Nugget!" are the types of headlines that can be found on the covers of Western and Eastern Treasures, a magazine for detector fans.

Searching a public beach for lost jewelry and coins is generally permitted under state laws, metal detecting in state and Federal parks and historical sites is not. People with detectors who hunt for Civil War and Revolutionary War relics (like uniform buttons and bullets) in landmark battlefields can disrupt these areas irrevocably if they start digging up what they detect, said Veletta Canouts, deputy program manager for archeology and ethnography for the National Parks Service.

"Unless you're working within some legitimate excavation planning process, there's little way of knowing the context of what you're digging up," Ms. Canouts said. "And once you destroy an archeological site, you can't put it back together again."

People who use detectors recreationally have to abide by Federal laws like the Archeological Resources Protection Act of 1979, which says any item found on public property is protected by the Government and is not the property of the finder. Some states protect items more than 100 years old found on private property.

The metal detecting community, which includes hundreds of regional organizations as well as the Federation of Metal Detectors and Archeological Clubs, a national group, discourages illegal searches. "There are a few bad apples out there who trespass and remove artifacts without permission," Mr. Nannetti said. "The best we can do is tell them we don't agree with what they're doing and try to get them to stop so that they don't spoil it for the rest of us."

Why the String Stays Bright After a Bulb Burns Out

The bulbs in a string of holiday lights are wired in series, which means that the current flows through each bulb, one after the other. Ordinarily, that would mean that if one bulb burned out, the circuit would be broken and none of the bulbs in that circuit would work. But each bulb contains a shunt that prevents that from happening.

CURRENT FLOW

Each light consists of a socket, a housing and a glass bulb. Wires from the string enter and exit the socket at the base, forming contacts for the two wires from the bulb, known as support wires.

BULB HOUSING

The support wires from the bulb pass through the housing and up the sides. So when the housing is seated in the socket, contact is made with the wires from the string.

THE FILAMENT AND THE SHUNT

At the end of the two support wires in the bulb is a filament, a fine wire that produces light when current passes through. Just below the filament is another wire called the shunt, which is wrapped around the support wires. The shunt is covered in a resistive coating so that ordinarily the electric current takes the path of least resistance, through the filament. If the filament burns out, the current tries to go through the shunt. The heat that is produced melts the resistive material, completing the circuit. So the current passes through the nonworking light and on to the next one.

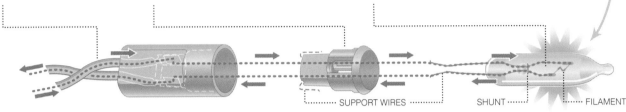

SUPPORT WIRES   SHUNT   FILAMENT

A Simple Chip Controls the Effect

The typical string of holiday lights is actually three strings, each on a separate circuit. The first bulb is on circuit No. 1, the second on circuit No. 2, the third on No. 3 and the fourth on No. 1 again, and the pattern repeats for the length of the string, typically 75 or 150 bulbs. If all three circuits are on at the same time, the entire string appears lit. If the circuits are switched on and off in a timed sequence, the lights appear to chase one another along the string, faster or slower depending upon the timing.

CIRCUIT BOARD

The board contains a small, inexpensive chip that controls the on-off timing of the three circuits. It also has diodes and other devices that reduce the voltage through the board from the normal 110 or 120 volts (which would quickly destroy the chip) to about 5 volts. The voltage to the three circuits remains 110 volts.

WEATHER-TIGHT SEAL

The plastic housing has a seal to keep moisture out. The tops of the screws that hold it together are coated with a waterproof compound.

CHIP

CONTROL KNOB

The knob operates a simple potentiometer, a variable resistor that changes the current entering the microprocessor.

Source: Najeh Rahman, Minami International Corporation

John Papasian

Lighting up the Holidays: On Chaser, on Dimmer, on Twinkler and Flasher

By Sally McGrane

The city of Taylor, Tex., started its holiday lighting display in the mid-1990s. More than 130,000 lights later, downtown Taylor is ablaze beginning on Thanksgiving weekend with brightly lighted Christmas trees, characters from nursery rhymes, a wagon train, teepees, a 20-foot sailing ship, a Santa Claus with his sleigh and eight reindeer, the Taj Mahal and a rocket ship.

This is no small project: Shanta Kuhl, executive vice president of the local Chamber of Commerce, said 300 volunteers spend about four weeks putting up lights in Taylor, a city of 15,000 northeast of Austin. "People just love it," Ms. Kuhl said.

These days holiday lights are taken for granted as almost every city in America puts up some type of display. And more than 80 million households in the United States use them to decorate each year, according to the Minami International Corporation, one of the largest manufacturers of holiday lights.

Today's holiday lights may be a far cry from the traditional candles that were once used to celebrate the winter solstice or, in the Christian faith, to represent Jesus as the light of the world.

The first recorded use of lights on a Christmas tree was in 1658 when candles were used in southern Germany, according to "Christmas Ornaments, Lights and Decorations: A Collector's Identification and Value Guide," by George Johnson. The candles were attached to trees by melting them directly onto the branches, attaching them with pins or clipping them on with special holders.

In the 1850's, gas tubes were added to real trees to provide lighting. In 1878, a British company unsuccessfully marketed an iron Christmas tree that contained channels for gas, which could be lighted through tiny holes in the branches.

The first electric Christmas lights were developed in 1882, shortly after the advent of commercial electricity. Each light had to be individually wired, however, which made them very expensive. By the 1920's, the cost of electric lights had dropped, and electricity was readily available.

Over the years, fashions in holiday lights have come and gone. In the early part of the century, lights of glass or plastic with figures of animals, fruits, Santas and even Uncle Sam were often used to decorate Christmas trees. In the 1950's, many people used bubble lights, which have colored liquid and a special chemical that sends bubbles up the glass when the light heats up. Rotating color wheels, twinkling lights and miniature lights followed.

Julie Ritzer Ross, who covers trends in Christmas lights for Selling Christmas Decorations magazine in Randolph, N.J., said clear or white lights were still popular for indoor lighting.

Retailers say they are selling more colored lights than in the past. And bubble lights have made a comeback.

Holiday lighting has also seen a large swing toward the outdoors. "People are getting increasingly competitive about outdoing their neighbors with outdoor decorating," Ms. Ritzer Ross said. "People are putting more lights outdoors — trees, bushes, hanging on eaves, wherever they can put them." Outdoors "icicles are still very big," she said.

While retailers might think that outdoor lighting fanatics are just trying to outdo their neighbors, they may have it wrong. For many enthusiasts, decorations are about Christmas cheer, no matter how much wattage is required.

Andy Jezioro, of Manassas, Va., decorates his house with about 40,000 lights. Mr. Jezioro, who takes 10 days off work around Thanksgiving to set up his display, explained: "It's kind of a childhood fantasy to have a house with lights on it — my parents would take us to see the houses with lights, and we never had the opportunity to decorate our house because we always rented."

Mr. Jezioro said the 400 strands of lights (which his wife, Barbara, tests before he hangs them) added about $350 to his electricity bill in December. He has also revamped his house's wiring. "All told with the electronics, I have about five or six thousand dollars invested," he said. When a display has at least 20,000 lights, it can be considered "very serious," he said.

Once the lights are up at the Jezioros' house, people start coming. "I love seeing the families in their cars with the kids and the blankets in the back seat," Mr. Jezioro said. "It makes it all worthwhile."

Blade Runner

A robotic lawnmower uses a combination of an electric guide wire and sensors to keep it on the job and out of flower beds and neighbors' yards. Here's a look inside a Robomow model, which can cut an average-sized lawn in a couple of hours.

WIRE SENSORS

Sensors in the mower detect a magnetic field around the guide wire. When the mower first starts cutting, it straddles the wire, following it around the perimeter of the lawn. The mower's electronics keep track of the turns the mower makes, and the device is designed to make up to two and a half trips around the perimeter before it starts mowing back and forth. The mower stops, backs up, turns and heads off at a slightly different angle each time its sensor detects the magnetic field so the entire lawn is eventually mowed.

MANUAL CONTROLLER

A radio controller can be lifted out of the mower's plastic shell for manual control of the mower outside the perimeter wire.

DRIVE MOTORS

Each of the two drive wheels has its own electric motor and gears. The mower turns by operating the wheels at different speeds or turning them in different directions.

FRONT WHEEL MOUNT

The front wheel swivels like a caster. Cutting height adjustments are made here and in the rear.

BLADE MOTOR

CIRCUIT BOARD

BATTERY PACK

The mower has two 12-volt sealed lead-acid batteries, connected in series to provide 24-volt power for the drive and blade motors.

GUIDE WIRE

A small current (from a 9-volt battery) running through the wire creates a magnetic field around it.

BUMPER SENSORS

Electric sensors in the bumpers are triggered when the mower hits an obstacle, causing it to back up and turn. The mower also has tilt sensors that will stop the machine if it heads down a steep slope.

THE BLADES

The mower has three blades, arranged in a triangle to cut a 22-inch swath. Each blade has its own motor, which operates at up to 5,800 revolutions per minute. When taller grass is encountered, the mower's electronic circuits sense the extra load on the motors and increase the power.

Source: Friendly Robotics

Frank O'Connell

Robotic Lawnmowers: For Homeowners Who Dream of Electric Sheep

By Mindy Sink

Although many futurists have envisioned robots that could relieve humans of ordinary tasks like cooking and cleaning, people must still do most of the domestic drudge work, at least inside the house. But outdoors, the future is here, in the form of mow-bots.

One of these devices is the Robomow, designed by Friendly Robotics, whose United States office is in Abilene, Tex. Robomow can propel itself around the yard, cutting the grass and even turning the clippings into mulch, eliminating raking from the to-do list.

Robomow is not the only robotic lawnmower around. Husqvarna, a Swedish manufacturer, makes two robotic mowers, including one that was introduced in 1995 and runs on solar power.

The lawnmowers are similar to another kind of household robot, machines that clean the bottom of swimming pools, in that they only work in a defined area. Outside of that area, these machines can't really cope. They may indeed be robotic, but they are a far cry from the autonomous, thinking, problem-solving machines of science fiction.

Robotic lawnmowers, for instance, need to be told where to go and, more important, where not to (like flower beds or the street). Guidance is offered by a wire that the homeowner installs around the perimeter of the yard (usually on the surface of the ground, although it can be buried a few inches deep). A low voltage current is passed through the wire, creating a radio-frequency signal; sensors in the mower detect the signal and keep the machine in the mowable area.

Although robotic mowers may appeal to anyone who needs to cut the grass, some people could benefit more than others from this hands-free approach.

"We have a lot of people who say, 'I have waited all my life for something like this,'" said John Bunton, products manager for Friendly Robotics. "People with allergies, disabilities, the elderly, people with heart disease — for the first time, a lot of these people can maintain their own lawn."

Mr. Bunton said his company's machines, which use battery power for both propulsion and cutting, were environmentally friendly, with none of the air pollution and little of the noise of gas-powered mowers.

The machines from both companies look a lot like canister vacuum cleaners without hoses or attachments. They tend to be fairly hefty: one Robomow model tips the scales at 71 pounds, including its rechargeable lead-acid batteries. The solar-powered mower from Husqvarna, however, weighs only 16 pounds.

After the Robomow mows along the perimeter one or two times, it starts cutting from one side of the lawn to the other, moving over for each new path. The Solar Mower and Auto Mower turn whenever they reach the perimeter, cutting swaths of grass in angular patterns that are not repeated until all the grass is cut.

Although the mowers have proximity sensors so they can avoid large obstacles, the lawn must be cleared of objects, just as it would be for an ordinary mower, before a robotic mower begins work.

"It's not going to go around and pick things up for you," Mr. Bunton said. "You have to make sure the hose is picked up, large limbs, toys, flip-flops." Human assistance also may be required if the mower encounters hills and tight corners.

The Robomow has a manual mode, but it is hardly the familiar push mowing of the past. The user simply follows the mower while holding a remote control keypad that is attached to the Robomow by a cord. That lets the user guide the mower without breaking a sweat.

The Sole of a New Sneaker

Sneakers that light up with each step are a perennial favorite among young children. The flashing effect is produced by a simple electronic mechanism in the sole that contains a battery, switch and controller chip.

FLASHING UNIT

LIGHT SYSTEM EMBEDDED IN THE SOLE OF THE SHOE

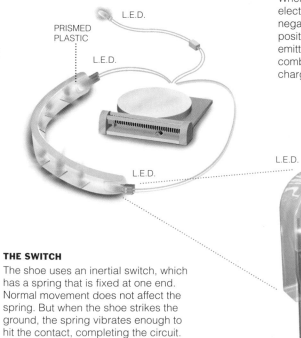

BATTERY   CIRCUIT BOARD   EMBEDDED CHIP

THE UNIT

The typical flashing sneaker uses a three-volt lithium battery. A child will usually outgrow the shoes before the batteries die. Electricity flows through a switch to an embedded chip, which controls the timing and pattern of flashes. When L.A. Gear produced a shoe for adults, the lighting system was in a removable cartridge so the battery could be replaced.

THE DISPLAY

In this model of shoe, two light-emitting diodes are inserted in a strip of clear plastic that wraps the light around the heel. Another wire leads to a diode behind clear plastic in the upper part of the shoe.

THE LIGHT-EMITTING DIODE

An L.E.D. is a semiconductor chip with negatively and positively charged areas that are separated by a junction. When voltage is applied, electrons flow from the negative side to the positive. Photons are emitted when the electrons combine with the positive charges.

SWITCH

CIRCUIT BOARD (WITH EMBEDDED CHIP)

BATTERY

SPRING   CONTACT

SPRING

CONTACT

CURRENT

PRISMED PLASTIC

L.E.D.

L.E.D.

L.E.D.

L.E.D.

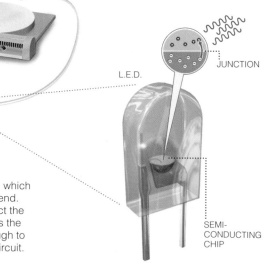

JUNCTION

L.E.D.

SEMI-CONDUCTING CHIP

THE SWITCH

The shoe uses an inertial switch, which has a spring that is fixed at one end. Normal movement does not affect the spring. But when the shoe strikes the ground, the spring vibrates enough to hit the contact, completing the circuit.

Mika Gröndahl and Cristina Rivero; Photographs by Tony Cenicola

Light on Their Feet: Flashy Footwear for Children

By Henry Fountain

Technological advances can be both good and bad, as Alfred Acree found out one night in April 1993.

Mr. Acree was being chased on foot in Charles City, Va., by police officers who suspected him of dealing in drugs. At the time, Mr. Acree was wearing the latest in footwear: a pair of running shoes from L.A. Gear with light-emitting diodes in the heels, designed to emit red flashes with each step. The shoes were designed to make runners more visible to drivers on dark roads, but on this night they made it easy for the police to spot Mr. Acree as he fled through the woods.

Good for the police, bad for Mr. Acree.

Millions of children have had happier experiences with sneakers that blink as they walk. Battery-powered children's sneakers were introduced by L.A. Gear in 1992 and were a sensation in their first three years, said Mark Goldston, a former chief executive of the company. At one point, L.A. Gear offered 30 to 40 different models. "We took the market by rage," Mr. Goldston said. "We sold millions of these things."

When L.A. Gear offered similar shoes for adults, the shoes did not fare as well. "It was a hot idea for six months, like many products in the adult market," Mr. Goldston said.

Sales of flashing children's shoes eventually dipped, too, but the children's market remains steady. Such shoes are now made by companies like Stride Rite and Skechers, as well as by L.A. Gear, and the red lights have been supplemented by green and other colors. Some models blink in patterns up and down the outside of the sole, some flash around the heel, and some even have diodes sewn into the uppers.

"They're like little red wagons — every kid has to have them," said Robert Greenberg, who created L.A. Gear in the 1980's and became a co-founder of Skechers in 1992.

Most, if not all, lighted sneakers use a system developed in the late 1980's by Nicholas Rodgers, a Canadian inventor. At the time, Mr. Rodgers was living in northern Ontario and wanted his young daughters to be more visible when they played outside on dark winter afternoons.

The solution he came up with, a ring of lights that fit over their boots, was clunky, he said, "but a lot of people thought it was pretty cool." So he worked for several years to refine the design and received his first patent in 1989. He sold the idea to a footwear wholesaler, who in turn took it to L.A. Gear.

Mr. Rodgers may have created the first marketable design, but inventors had struggled for decades to marry lights and footwear. Often the intention was safety: a lighted shoe makes the wearer easier to see. But just as often, the goal was simple novelty or entertainment. A 1979 patent for "flashing disco shoes," for example, featured lights embedded in a clear plastic shoe with four-inch heels.

Patents for lighted shoes with an on-off switch in the sole or heel were issued at least as far back as 1933, but early designs suffered from a common drawback, as described in a later patent: the battery quickly went dead when the lights were continuously on, "such as occurs in activities involving much standing."

Various switch designs were developed to get around this problem. One used a ball bearing that rolled in a tube as the foot moved, completing an electrical circuit when it reached contacts at the tube's end.

In its first models, L.A. Gear used a mercury switch. But mercury is toxic, and when environmental regulators in Minnesota learned in 1994 that thousands of children were walking around with it in their shoes, they took action. The State Legislature voted to ban the sale of the shoes, and L.A. Gear agreed to pay to dispose of the roughly 20,000 pairs of sneakers it had sold in the state.

Similar bans were considered in other states, but L.A. Gear had shifted by then to a different switch, the kind used in most flashing shoes sold today. Called an inertial switch, it consists of a metal spring that is connected to the electrical source on one end; at rest, it is close to, but not touching, a metal contact at the other end. If the shoe moves violently or hits the ground when the wearer takes a step, the spring vibrates enough to touch the contact, completing the circuit.

Rover's In for a Shock

An electronic fence, used in combination with training, discourages dogs from straying off their property. The fence has two components: a wire loop that is buried around the perimeter of the yard and transmits a low-power signal, and a collar containing a receiver and electrodes for delivering an unpleasant shock to the dog when it gets too close to the wire.

THE UNDERGROUND FENCE

A wire loop is buried about two inches deep. It is connected to the transmitter and emits radio signals that can be picked up by the dog's collar. In some systems, a second loop can be used to protect things like a pool or a tree.

Interior areas can be protected by small loops. The twisted-wire segment does not activate the collar.

Inside the collar

The collar, which can weigh as little as 1.5 ounces, picks up the radio signals emitted by the buried wire when the dog gets too close. The collar starts beeping to warn the dog. A dog that tries to cross the boundary anyway gets a painful but brief shock, which is considered safe. The intensity of the shock can be adjusted on most collars, and there are more powerful collars for more determined dogs.

ON ALERT

Most transmitters send out an AM signal in the range of approximately 7 to 10 kilohertz. The transmitters are protected against lightning strikes.

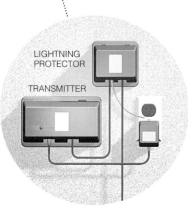

LIGHTNING PROTECTOR

TRANSMITTER

Setting boundaries in the suburbs

An electronic fence is generally less expensive than a real fence (a four-foot chain-link fence for a quarter-acre would cost about $2,000, including installation). For a dog that is accustomed to the fence, the threat alone suffices. But some dogs never learn.

A sneeze, not a yelp

The Spray Control Fence from PetSafe uses a collar that substitutes a spray of citronella, a sharp-smelling oil that dogs find unpleasant, for a shock.

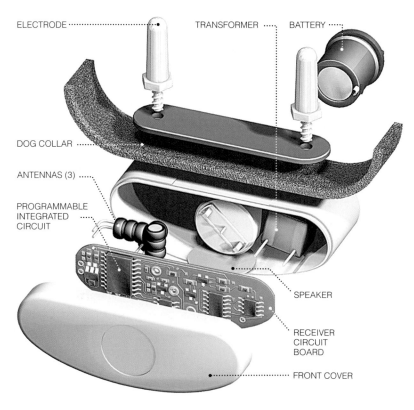

ELECTRODE

TRANSFORMER

BATTERY

DOG COLLAR

ANTENNAS (3)

PROGRAMMABLE INTEGRATED CIRCUIT

SPEAKER

RECEIVER CIRCUIT BOARD

FRONT COVER

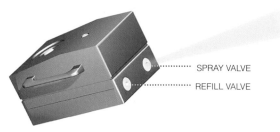

SPRAY VALVE

REFILL VALVE

Sources: ISCO Enterprises; Radio Systems Corp.

Frank O'Connell

Hidden Fences Let the Family Dog Know Its Limits

By Shelly Freierman

No amount of scolding, pleading or foot-stomping will keep most dogs home in their yards and out of the streets. So some frazzled owners try to turn their neighborhood travelers into docile homebodies with radio-controlled fences.

These hidden fences consist of wire loops, buried an inch or two below the surface, that emit radio signals when connected to transmitters. Each dog wears a collar with a receiver; when the dog gets too close to the wire, the collar emits a warning beep. If the dog ignores the beep and gets closer still, the collar's two metal prongs give the dog a mild shock.

Debby and Ed Christian have two Old English sheepdogs who are so well trained by a hidden fence that they never leave the Christians' yard in Acton, Mass. "I could probably not even use the collars anymore," Debby Christian said, "and the dogs would stay. Even when another dog comes into the yard, my dogs won't leave."

The Christians bought a version called Invisible Fence, made by ISCO Enterprises. The dogs, Fletcher and Phoebe, figured out the invisible barrier right away.

When dogs are trained, flags are posted along the boundary wire and the dogs are walked toward them on a leash. As the dogs approach the flags, the dogs hear their collars beep. They are told "no" and are turned back toward the house. The final step in the process, which is repeated over several days, is to deliver a mild shock — or correction, the euphemism favored by all the hidden-fence manufacturers — to reinforce the idea that the dog should not cross the boundary line. Some dogs require only one shock to learn to stay within the hidden fence, while others need several.

The shock hurts the dog, but it causes no lasting harm. The dogs yelp, said Melissa Bain, a resident in clinical animal behavior at the School of Veterinary Medicine at the University of California at Davis, but the punishment has to be unpleasant to be effective.

"The shock works because it is simple aversion learning," said Katherine Houpt, the director of the Animal Behavioral Clinic for the College of Veterinary Medicine at Cornell University. "However, it works best with those dogs that are not highly motivated to leave." Among the more difficult pupils she described are male dogs that are not neutered, dogs that like to hunt or chase cars and highly territorial dogs that are eager to cross boundaries to scare strangers.

The Invisible Fence was invented in 1973 by Richard Peck, a traveling salesman and animal lover who was saddened by the number of dead dogs he saw on the side of the road. When his patents expired in 1990, several other companies joined the industry. Some of the products, like Invisible Fence and DogWatch, made by DogWatch Inc., are sold by independent dealers who install the product and usually train the dogs (and owners). Even the most expensive hidden fence is considerably cheaper than a chain-link or wood fence.

A hidden fence may protect a dog that is prone to roam, but it does not necessarily protect strangers approaching the yard. The dog knows where the boundary is, Dr. Houpt said, but a visitor like a letter carrier may not know that there is a fence and that the dog's territory is being invaded.

Dr. Bain also says that is a drawback of hidden fences. If a stranger steps into the yard, she said, the dog can become aggressive. "The aggression can range from chasing them to biting them, depending on the dog," she said. "We tell people with aggressive dogs to get a real fence."

Some dogs can outsmart a hidden fence. They have figured out that if they stand near the boundary wire to trigger the beep, the battery will eventually run out and they can come and go without a shock. One solution designed to discourage escapees is a collar that progressively increases the intensity of the shock as the dog approaches the fence.

Still, no matter what the companies come up with, some dogs won't stay put — a cat across the street can become just too enticing. "It is always better to have the dog in the house unless you can walk it on a leash," Dr. Houpt said. "Yes, it's true a dog needs exercise, but probably so do its owners."

Not Your Average Household Pet

Sony's Aibo robot pet is a walking, talking computer (without the keyboard and the hard drive). It is about 10 inches high and 11 inches from head to tail, weighs slightly more than 3 pounds and, when fitted with the proper software, will evolve from puppyhood to adulthood right before your eyes. But even a fully trained adult Aibo won't fetch the slippers.

THE HEAD

Aibo's head contains touch sensors on top and on the chin, an infrared sensor for measuring distance, stereo microphones and a speaker, an image sensor and light-emitting diodes, or L.E.D.s, that signal the robot's condition or mood. The head moves in several directions, the mouth opens and closes; even the ears are motorized. Aibo is instructed through voice commands or by pressing the various touch sensors.

INSIDE THE BODY

The robot's brains are housed in a black plastic box slightly larger than a portable tape player. The main circuit board includes a 200-megahertz processor, two 16-megabyte RAM chips and 4 megabytes of read-only memory. Other chips include an input-output controller and two of Sony's own design. A rechargeable lithium-ion battery supplies the power. The body also contains a heat sensor, a vibration sensor and an accelerometer, among other components.

BACK AND TAIL

A small motor moves the tail, and L.E.D.'s light up and blink, signaling the robot's moods. The back contains another touch sensor.

UPGRADES

Additional software that enables the robot to be more active, learn more voice commands and perform simple tricks is supplied on removable flash memory cards. The main box also has room for a wireless network card so that the robot can be controlled by a computer. Aibo uses a software platform called Open-R, which was developed by Sony.

L.E.D.'S

IMAGE SENSOR
A 100,000-pixel CMOS chip allows the robot to sense colors, particularly pink, green and blue. Aibo can store up to seven images, which can be downloaded to a PC. The robot uses information from the image sensor in combination with distance data from the infrared sensor to seek and follow the pink ball.

TOUCH SENSOR

STEREO MICROPHONE

TOUCH SENSOR

L.E.D.'S

RAM

PROCESSOR

ON-OFF BUTTON

STATUS LIGHT
Changes color according to activity.

MEMORY CARD

WIRELESS NETWORK

LEGS

A motor in each knee and two in each shoulder allow the robot to walk, sit, make gestures and kick the ball. Pressure sensors in the bottom of the feet detect when the robot is picked up; when this happens, the joints "relax," allowing Aibo to be cuddled.

MOTOR

Frank O'Connell

This Robot's So Lifelike, It's Almost Canine

By David Pogue

Science fiction writers of the 1950's might have been astounded to find that the world's first mass-produced, widely adopted robot, Sony's second-generation Aibo ERS-210, did not do housework, fetch slippers or aspire to dominate mankind.

In fact, it does little more than play on the carpet. Nonetheless, this new machine is among the most impressive pieces of circuitry to come along in years.

Most people think of this genuine, walking, autonomous robot as a dog, but it has the size, weight and moodiness of a cat. It responds to the name Aibo (EYE-bo), which is Japanese for "companion" or "pal."

(Sony said Aibo also stood for Artificial Intelligence Robot. Well, maybe if you're really tired.)

Clad in metallic black, silver or gold plastic, this virtual pet resembles an armored terrier in a space helmet. It has 32 megabytes of memory and a 90-minute battery.

Twenty motors provide realistic if slightly arthritic motion to the leg joints, tail, ears, head, neck and jaw. A 32-bit processor and 32 megabytes of memory serve as its brain. It has a color camera for eyes, stereo microphones for ears, touch sensors, a synthesizer for sound and internal gyroscopes for balance. It even has an infrared transceiver in its muzzle to help it avoid slamming into a wall, which is always an embarrassing moment for a robot.

The software that drives this puppy is sophisticated, constantly topping itself by offering new tricks. Through cleverly programmed body language, Aibo does an effective impersonation of a cute and frisky puppy. For example, Aibo exhibits what its makers call emotions — six different ones — responds to praise and punishment and shows off a never-ending array of doggy gestures, from yawning and scratching to lifting a leg. It expresses moods visually (its eyes and tail light up in different colors) and sonically, using melodies and chirps that make R2-D2 from "Star Wars" sound noncommunicative.

Aibo's best trick is eyeing, chasing and then kicking a pink plastic ball (which comes with the dog), although it responds with equal enthusiasm to almost anything made of red or pink plastic.

Over time, each Aibo can develop a personality; in fact, Sony said that at one Aibo get-together, owners were able to distinguish their pets from other Aibo dogs.

"Take a picture" is one of about 40 voice commands Aibo understands; the others include standard commands like "Sit," "Shake" and "Lie down." But that's just the beginning. Saying "Let's play" triggers a round of mimic the master, in which the dog's robotic whistles hilariously duplicates the rise and fall of the speaker's voice.

But Aibo doesn't always respond. Sony says these displays of attitude heighten the robot's realism, although in this case "attitude" is scarcely distinguishable from "not working."

Some Aibo buyers are apartment dwellers without the time or permission to care for real dogs; they appreciate a pet that seems to offer the affection of a live dog but few of the drawbacks. But most Aibo buyers are not looking for an alternative to a "fuzzy dog" (as Aibo devotees call real pets). Instead, they seem to be driven by something far more emotional — parental instinct — a phenomenon first visible in 2000 when an early, limited-edition (and limited-function) model went on sale. Sony sold 3,000 of them in the first half hour, and today these original Aibos can fetch astronomical prices.

Aibo owners are a decidedly passionate lot; they have their own Web sites, conduct weekly Internet chats and even hold regional get-togethers. Their bulletin boards are filled with eyebrow-raising commentary. Some other sites offer hacker tools for extracting data from the pet's Memory Stick.

Sony even provides a Windows software kit that that allows Aibo fans to program new movements, sounds and melodies, which can then be shared online. The kit also allows Aibo to be commanded from a PC by remote control. So you can teach this new dog even newer tricks.

Frank Sinatra He Isn't

It's a question that's been asked countless times: If they can send a man to the moon, why can't they come up with a rubber-and-plastic mounted fish that thumps its tail and lip-syncs "Take Me to the River"? Well they can, and they did. Here's a look inside Big Mouth Billy Bass.

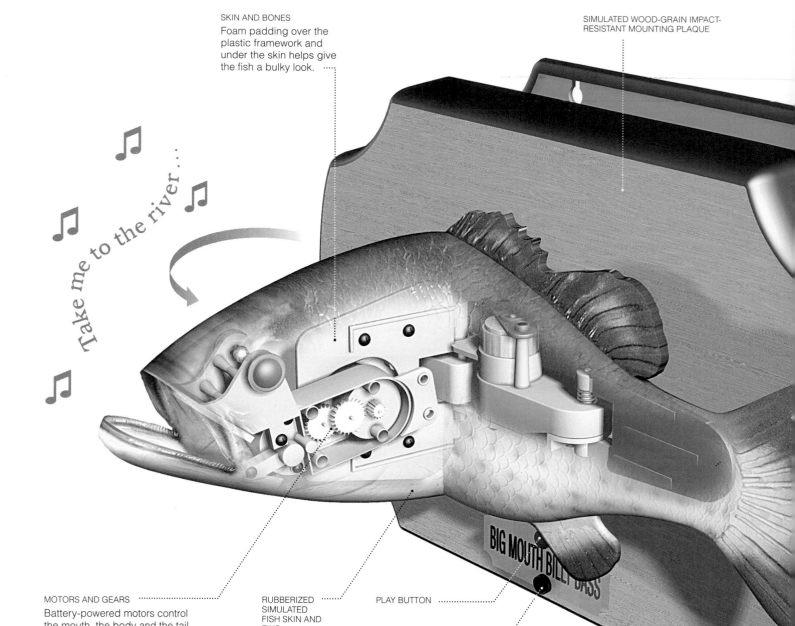

SKIN AND BONES
Foam padding over the plastic framework and under the skin helps give the fish a bulky look.

SIMULATED WOOD-GRAIN IMPACT-RESISTANT MOUNTING PLAQUE

Take me to the river...

MOTORS AND GEARS
Battery-powered motors control the mouth, the body and the tail. All are connected via plastic gears to mechanisms that move one way when the motor operates and snap back to the original position, aided by a spring, when the motor shuts off. The body mechanism, for example, swings the front of the fish out from the plaque when the singing starts.

RUBBERIZED SIMULATED FISH SKIN AND FINS

PLAY BUTTON

MOTION SENSOR
A photocell acts as a motion detector, tripping when someone moves in front of it and blocks the light. That sends a signal to the circuit board to turn the fish on. The fish can also be operated by pushing the play button.

BIG MOUTH BILLY BASS

Source: Gemmy Industries Corp.

Frank O'Connell

A Trophy Fish That Quickly Gets Under the Skin

By Heidi Schuessler

SPEAKER

CIRCUIT BOARD
A small circuit board includes a chip, embedded in resin, that contains digitized music and instructions to control the motors.

Gag gifts generally have short half-lives. For all the publicity generated by Pet Rocks back in the 1970s, for example, for such a product to last more than a season or two is a rarity.

In the gag gift universe, 2000 was the year of Big Mouth Billy Bass, an eerily realistic, and multitalented, 13-inch rubber and plastic large-mouth bass mounted on a trophy plaque. Push a button or trigger a motion sensor and the fish moves its head, tail and mouth while lip-syncing to songs like "Don't Worry, Be Happy" or "Take Me to the River."

Billy Bass actually contained some sophisticated circuitry that enabled it to do what it did. But fundamentally it was a one-trick pony, designed to amuse and astound. People loved it — for a while.

When Ed Caldwell, an engineer, saw a Big Mouth Billy Bass at his father's house, he got an idea. He borrowed the fake fish and took it to his office at Cymbolic Sciences in Bellingham, Wash. "I put it on my co-workers' desks early in the morning," he said. "When they came in it would start singing and wiggling around — it scared them to death, but they thought it was hysterical."

The gag gift first appeared at sporting goods stores and was an instant hit. At the Bass Pro Shops Outdoor World in Grapevine, Tex., 500 units sold in under an hour. "I had friends calling me to ask if I could sneak them higher up on the waiting list," said Rachel Rickman, the store's events and promotions manager.

It didn't take long for the fish to reel in more than anglers. Billy Bass soon appeared in regular department stores and toy stores. Queen Elizabeth was even said to have one, displayed on the grand piano at Balmoral Castle.

"It transcends class because it is so off the scale in terms of taste," said Dr. Robert Thompson, professor of media and pop culture at Syracuse University and president of the Popular Culture Association. Billy Bass, he added, is part of a proud tradition of gag gifts that goes back 120 years. "There's something incredibly American about Billy," he said. "He's the best cheesy gift we've ever come up with."

How did anyone ever think to combine pop tunes and taxidermy? Call it a fluke: In 1998, Joe and Barbara Pellettieri drove past that same Bass Pro Shops store in Grapevine while brainstorming ideas for new products. Mrs. Pellettieri looked at the bass in the store's logo and said, "What about a singing fish?"

It seemed a good fit for Gemmy Industries, an animated-novelty company in Irving, Tex. Mr. Pellettieri is the company's vice president for product development. He set to work on a prototype but it almost died on the drawing board. "It was hideous," said Jim Van den Dyssel, Gemmy's vice president for marketing. "The fish looked prehistoric."

Mr. Pellettieri didn't give up. "It's not an exact science," he said. "It's more a gut-instinct thing." Another prototype was developed, but it simply flopped around, uninspired. When he asked the engineers to add a third motor to make the fish's head turn at a right angle to the plaque, he knew he had a winner.

"You look at it and the intelligent side of your brain says, 'What is the world coming to?'" Dr. Thompson said. "Then the other side of your brain says, 'Cool — that fish just looked at me.'"

But Can It Dance?

The singing greeting card is a mixture of low and high technology. As to whether it will ever take over the greeting card market: Que sera, sera.

THE CARD breaks into song, and sometimes a message, too (recorded by the card company or by the sender), when you open it. The movement pulls back a plastic tab that separates two electrical connectors, allowing them to meet and complete a circuit. Then the sound chip and speaker play what is in the chip's memory.

OPENING THE CARD PULLS OUT THE TAB . . .

. . . AND THE CIRCUIT IS COMPLETED.

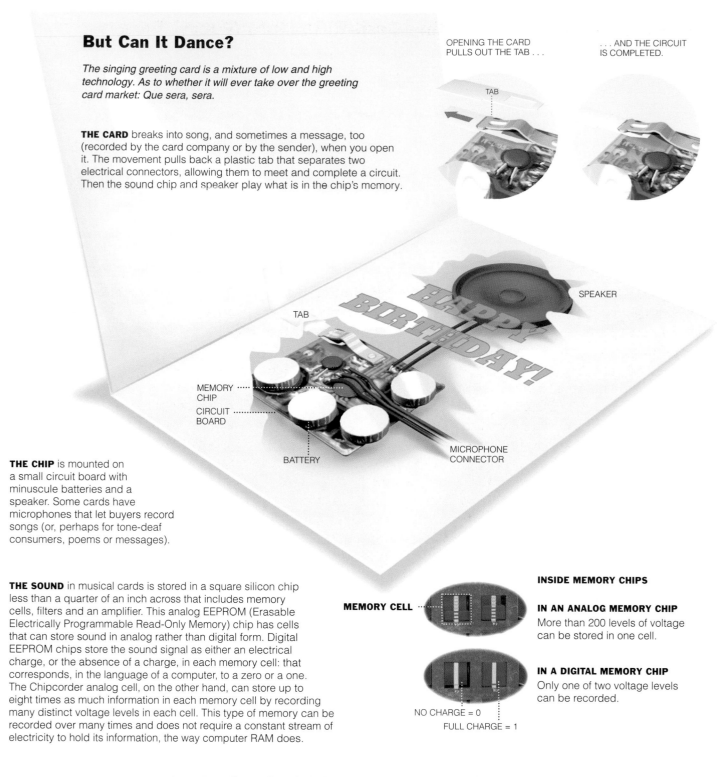

TAB

SPEAKER

TAB

MEMORY CHIP

CIRCUIT BOARD

BATTERY

MICROPHONE CONNECTOR

HAPPY BIRTHDAY!

THE CHIP is mounted on a small circuit board with minuscule batteries and a speaker. Some cards have microphones that let buyers record songs (or, perhaps for tone-deaf consumers, poems or messages).

THE SOUND in musical cards is stored in a square silicon chip less than a quarter of an inch across that includes memory cells, filters and an amplifier. This analog EEPROM (Erasable Electrically Programmable Read-Only Memory) chip has cells that can store sound in analog rather than digital form. Digital EEPROM chips store the sound signal as either an electrical charge, or the absence of a charge, in each memory cell: that corresponds, in the language of a computer, to a zero or a one. The Chipcorder analog cell, on the other hand, can store up to eight times as much information in each memory cell by recording many distinct voltage levels in each cell. This type of memory can be recorded over many times and does not require a constant stream of electricity to hold its information, the way computer RAM does.

MEMORY CELL

NO CHARGE = 0

FULL CHARGE = 1

INSIDE MEMORY CHIPS

IN AN ANALOG MEMORY CHIP
More than 200 levels of voltage can be stored in one cell.

IN A DIGITAL MEMORY CHIP
Only one of two voltage levels can be recorded.

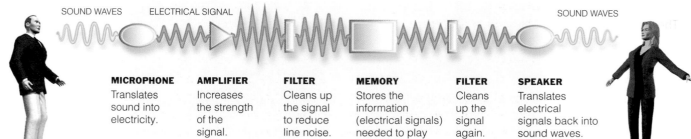

SOUND WAVES   ELECTRICAL SIGNAL   SOUND WAVES

MICROPHONE
Translates sound into electricity.

AMPLIFIER
Increases the strength of the signal.

FILTER
Cleans up the signal to reduce line noise.

MEMORY
Stores the information (electrical signals) needed to play back the sound.

FILTER
Cleans up the signal again.

SPEAKER
Translates electrical signals back into sound waves.

Mika Gröndahl

How a Chip Makes a Greeting Card Burst Into (Fuzzy) Song

By Matt Lake

On the outside, it looks like a regular greeting card. A little bulky, perhaps, and a little heavier than you'd expect, but otherwise, nothing out of the ordinary. But when you open it, a tune begins and a movie-preview voice booms out a message. Yes, this is a singing, talking greeting card.

Although cards like these have been around for decades, they are still enough of a novelty to surprise. And some of them even let senders record their own songs or messages.

The technology that underlies such cards is a mixture of different sciences: microelectronics, old-fashioned mechanics and economics. On a novelty item like this, price is of the essence. While people will spring for $3 for an ordinary card, they won't go much above that for one that sings. Fortunately, an inexpensive sound chip does most of the work. A leading chip, the tiny Chipcorder (from Information Storage Devices), not only provides enough room to record a short message or jingle, but also integrates sound filters and amplifiers — all on a silicon chip so small that six of them could fit comfortably on a postage stamp.

Chipcorders record analog sound, not the digital sound usually associated with microelectronics, and for good reasons. For one, it cuts out the expense of the two converters needed by computer sound cards and modems to convert analog signals into digital signals, then reverse the process. Since analog sound does not need to be encoded and compressed like digital sound, it takes up much less storage space, which also keeps costs down.

These sound-recording chips were first marketed in the early 1990's for use in telephone answering machines, portable dictation devices and interactive books. But since chips with relatively short recording times of 10 or 20 seconds can be manufactured quite cheaply, they migrated naturally to novelty items like greeting cards. At the beginning of 1998, they also found their way into more than 12,000 New York City taxicabs: They are the technology behind the celebrity voices that remind passengers to buckle up, pick up a receipt and take their belongings when they leave.

Regardless of what they are used for, Chipcorders need a signal to tell them to start playing. That is the only mechanical part of the whole operation, and it is very low tech. In greeting cards, the circuit board containing the Chipcorder is glued to a paper mounting. A plastic tab keeps two electrical contacts from touching each other and closing a circuit.

The paper mounting attached to the end of the plastic tab is then glued over the fold in the greeting card. Opening the card draws the tab back; it slips out from between the two contacts, allowing the circuit to be completed. The resulting electrical signal tells the Chipcorder to play. (In taxicabs, the Chipcorders' play signal comes from turning the meter on or off.)

The chips can easily be recorded over, for at least tens of thousands of repeat recordings. But not all singing greeting cards allow you to record your own messages, because it costs more to include that capability. All Chipcorders have the capacity for re-recording, but for customers to record their own sounds requires that each card have an additional switch and a microphone, and those cost money to buy and install.

It's economics that has prevented audio greeting cards from becoming widespread. The typical outlets for greeting cards — bookstores, drugstores, card shops and, increasingly, supermarkets — are high-overhead enterprises that pay close attention to profit margins. The Chipcorder, batteries and speaker alone can cost what a retail outlet is prepared to pay for a card, pricing singing cards out of the retail channel. So some big card companies that experimented with singing cards, like Gibson Greetings, no longer manufacture them.

Computers
and the
Internet

Analog Sound

Analog refers to information, like sound waves, that has been converted into an electronic pattern — in other words, into a pattern that is analogous to the original sound wave. The information can be stored or transmitted in this electronic form and then converted back to sound on the receiving end.

IN AN ANALOG PHONE CALL

1 Speaking into a telephone sends sound waves into the microphone in the mouthpiece.

2 The microphone is covered by a membrane that vibrates with the sound waves.

3 As the membrane vibrates it makes a magnet vibrate within an electrified wire coil. The motion within the coil creates a continuous but varying charge, or voltage, which is sent across a copper wire.

4 On the receiving end, the fluctuating voltage moves a magnet, which is attached to the diaphragm in the phone's earpiece speaker.

5 The vibrating membrane in the speaker converts the electrical pattern back into sound waves.

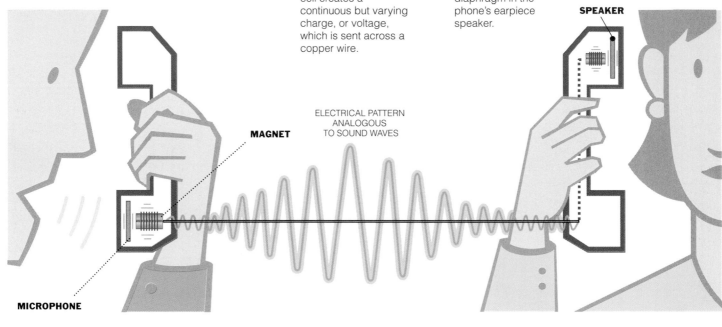

SPEAKER

ELECTRICAL PATTERN ANALOGOUS TO SOUND WAVES

MAGNET

MICROPHONE

Digital Sound

Digital refers to information converted into an electronic pattern that can be represented in a computer code as numbers, or digits. This code uses a simple binary system of 0s and 1s.

IN A DIGITAL PHONE CALL

1 At first, the process is the same in digital as in analog: the voice is picked up by the microphone in the phone's mouthpiece.

2 From the home telephone, the analog signal makes its way over the phone company's copper wires.

3 At the phone company's switching station, or a local conversion station near the home, the signal is run through a computer chip, which converts the electrical pattern to digital form. It can then be sent over a high-capacity fiber-optic line.

4 The chip measures, or samples, each wave at multiple points, sampling several thousand times each second.

5 Each measurement is then converted into a string of eight 1's or 0's.

6 Before leaving the phone company's portion of the network, the digital signal is converted back into analog form.

7 The speaker in the phone then converts the analog signal into audible sound waves.

ELECTRICAL PATTERN ANALOGOUS TO SOUND WAVES

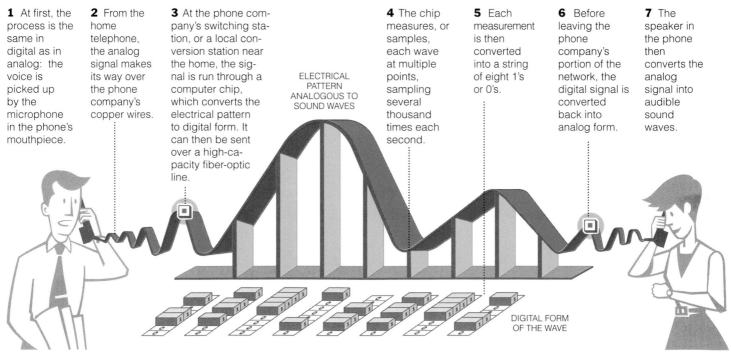

DIGITAL FORM OF THE WAVE

Gorka Sampedro and Kris Goodfellow

What Do They Mean by Digital, Anyhow?

By Tim Race

The digital age, they call it — this era in which information increasingly gets stored, retrieved and transmitted in an electronic form that computers can handle. But what does "digital" mean, anyhow?

To understand digital, it helps to be familiar with its technical precursor, analog. A basic example of analog technology is the conventional telephone, in which sound waves are converted into a fluctuating electrical pattern analogous to the original sound waves.

This principle — analogous electrical patterns — holds true whether sound is traveling over wires or via radio frequencies. It is also the basic concept behind television, for instance, although in television, the microphone and speaker are replaced by components that translate light waves into an electrical pattern and then back into light waves.

In the case of digital technology, information detectable by human ears or eyes is converted into an electronic pattern expressible as the digits of computer code. This code typically uses a simple binary system of zeroes and ones, which can also be expressed as "on" and "off" in an electrical current or, in the case of fiber optics, as "on" and "off" pulses of light.

Information can be captured and reproduced digitally with much more precision than with analog technology, and it can be compressed and stored in a much smaller physical format. And unlike analog information, digital information can be copied an infinite number of times without losing the fidelity or quality of the original version. With the ascent of the computer in recent decades has come the impulse to digitize information whenever possible. Telephone networks began incorporating digital techniques in the late 1970's, the audio CD introduced digital sound to consumer electronics in the early 80's, and CD-ROM's and the Web have brought sound and video to personal computers in the 90's. Television is now in a digital rollout that will continue until the end of the decade and beyond.

Slicing up the signal

An analog signal can be converted into a digital one by sampling — measuring the amplitude of the analog signal at regular intervals and representing those measurements in binary form, a string of 0s and 1s. If you think of an analog signal as a typical sine wave, sampling is like cutting the wave into uniform slices and measuring the height of each slice.

The more frequently a signal is sampled (the thinner the slices), the more accurately the digital signal will represent the analog one. If the signal is sampled once a second, for example, it won't be very accurate — it will miss all the information being conveyed during the time between samples. But sampling a lot more often — 20,000 times a second, say — will make the digital signal a much closer approximation of the analog signal.

It will always remain an approximation, but if the sampling rate is high enough the difference will be negligible. Sampling an old analog sound recording at a high rate, for example, will produce a digital version that most people won't be able to tell apart from the original (though there will always be purists who will complain that digitizing takes the warmth out of analog recordings).

The more you sample, of course, the more data you create, and that's why the digital world owes so much to the idea of compression. Unless you can compress some of the data in a digital video signal, for example, you'll end up with files that are so large they'll overflow your hard drive. And they'll be so bulky and unwieldy that they'll move through your computer like molasses.

A Layer Cake in Silicon

At the heart of every computer lies the silicon chip, the basic building block of which is the transistor. Transistors and other chip components are made by building up layers of silicon that have been "doped" with other elements to affect conductivity. Some areas are etched away and metal is deposited for wires. The result is a structure that performs electrically as designed.

Inside a Transistor

DOPING SILICON The N-type silicon layer has atoms like phosphorus, with extra electrons, added to make the general charge more negative. The P-type silicon has atoms like boron, with fewer electrons, making the charge in that region more positive.

THE GATE When there is no charge on the gate wire, no current flows between the source and the drain. A small charge applied to the gate wire, however, attracts electrons to an area of the P-type silicon, allowing current to flow

SIZE MATTERS Chip manufacturers are always working to make the wires thinner and shorter. Wire size is a primary governing factor in chip speed; shorter wires mean more transistors per chip, and signals travel shorter distances.

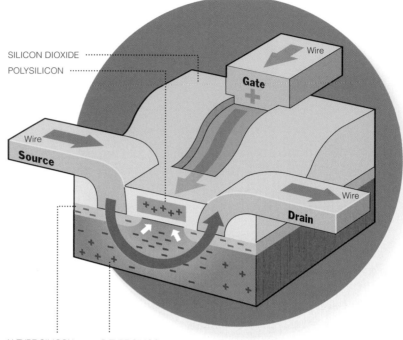

SILICON DIOXIDE
POLYSILICON
Wire
Gate
Source
Wire
Wire
Drain
N-TYPE SILICON
P-TYPE SILICON

Inside a chip

A SEA OF TRANSISTORS Chips can contain millions of transistors. The average chip is about the size of a fingernail, though it is usually contained in a larger piece of plastic. This chip has 7.5 million transistors.

CATEGORIES Chips are usually divided into general categories, like central processing units, memory chips and glue logic, which helps chips interact. But most C.P.U.s have plenty of memory on the chip, and some memory chips come with logic for computation. Glue logic are generally simple computational chips.

THE CORE At the core of a central processing chip is the arithmetic logic unit or units. These do basic operations like addition, multiplication and division. They can go faster if they have more transistors allocated to them.

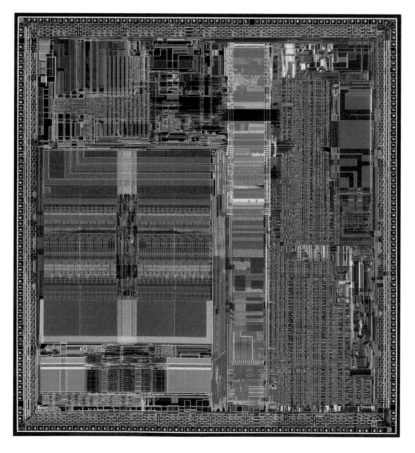

THE CACHE is another component common to processing chips. The size of a C.P.U.'s cache is important in determining how fast the C.P.U. will work. If data are kept in the local memory on the C.P.U.'s own chip, the C.P.U. can gain access to it very quickly. On the other hand, if a chunk of data is kept, or cached, elsewhere, the C.P.U. has to go looking for it. Adding more cache space on the C.P.U. chip is one of the most important ways to speed up chips.

THE CLOCK Most C.P.U.s come with a clock that sends out a signal pulse that acts like a metronome, helping to synchronize all the parts of the chip.

Gorka Sampedro and Mika Gröndahl

Behind the Curtain in a Digital Oz, the Simple Chip

By Peter Wayner

Deep in the heart of every computer is the silicon chip. The chip, with a snappy name and a sexy mystique, has a dark secret that seems almost scandalous next to the flashy data it throws onto the screen: A chip is a boring device that does what it's told. It's a mechanical automaton.

Chips are complex mechanisms built out of millions of simple switches. Each switch does what it is told and tells other switches what to do. When the orders have been executed, the chip produces an answer before pausing and doing it all again.

The basic switch at the heart of the chip is a transistor, a device made of silicon, a semiconductor. The silicon is called semiconducting because it can either conduct or not conduct electricity, depending on whether an electrical signal is present. In essence, the transistor is just an electrically operated switch.

When millions of transistors are joined, they can start doing useful things. The modern computer chip can do basic arithmetic and move data around in memory. That may not sound like much, but computer programmers have figured out how to string together millions of these simple arithmetic operations to do things like draw monsters or pass on jokes.

Not much has changed in the chip world since the early 70's, when many electronics companies figured out how to put several transistors on the same chip. Each year, the engineers find ways to create smaller transistors.

Smaller transistors help in two ways. First, more transistors can be packed on a chip. More transistors can often work faster because they can attack problems in parallel — dividing up the pieces of a problem, then combining the results for a final answer. Second, and just as important, signals travel shorter routes when transistors are packed closely together. Shorter routes mean faster chips, which mean faster computers.

The improvement in transistor manufacturing was envisioned by what is often known as Moore's Law, a sort of self-fulfilling prophecy. The rule, named after its formulator, Gordon Moore, the co-founder of the chip giant Intel, states that the speed and power of computer chips will double every 18 months.

Although the pace of change has sometimes been faster than the rule suggests, its prediction has been startlingly accurate. That may be because everyone in Silicon Valley believes it to be true and uses it as a benchmark for projects.

Robert Dreyer, a computer architect at Intel, said, "When we wonder about adding a feature or not, we ask, 'Does it move you off Moore's law?' People continue to predict that chip manufacturers will hit some wall, or physical law, that will stop the pace of innovation, but someone has always found an innovation that allows development to stay on schedule."

Many people in the industry regard the width of the basic wires on the silicon as a governing factor. The width that was once regarded as the minimum is one micron. (A human hair is about 100 microns wide.) That old minimum is now a quarter-micron, and many experts expect it to continue to shrink as well.

The Information Assembly Line

The first thing consumers are likely to see in a computer advertisement is some reference to a blazingly fast machine for sale that rips through games and spreadsheets at 700 or 800 MHz. That's "megahertz," a reference to the speed of the computer's clock switch. But a computer with a clock speed that is twice that of another is not necessarily twice as fast. Other factors enter into the equation, like the availability of the data being used in the calculation.

THE CONVEYOR BELT OF DATA

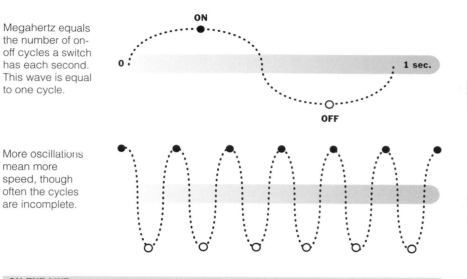

Megahertz equals the number of on-off cycles a switch has each second. This wave is equal to one cycle.

More oscillations mean more speed, though often the cycles are incomplete.

ON THE LINE

A computer processor has a clock that regulates the flow of information through the chip like a supervisor monitoring an assembly line. At each station on the line, data are assembled. Each time a clock signal oscillates, data move on to the next station — if all the data are available. If not, it really doesn't matter how fast the clock speed is. The processor will be kept waiting until the data arrive.

DATA WAREHOUSE

SPARE PARTS

FINAL PRODUCTS

1

Pieces of data are gathered and assembled at a very high speed.

2

At each station on the line, new data are added and are sent on to the next stop.

3

The line moves very quickly as long as all data needed are available.

4

If any piece of data needed for the end product is missing, the line must wait until it is found. That may require searching in a distant location and may slow the line.

5

The final product — the solution to a math problem, for example — is sent out (to memory) when all tasks are completed.

SPARE PART

John Papasian

Clock Speed and Megahertz: It's Really a Matter of Timing

By Peter Wayner

In the movie comedy "Spinal Tap," a rock guitarist explains to an interviewer that his amplifier is much better than others because the top reading on his dials is 11. When he wants to push his music to the limit, he can turn the dials to 11, while other, lesser guitarists get stuck at 10.

The scene is a favorite among engineers because it shows in a farcical way how numbers can produce a convincing and appealing illusion just as effectively as trompe l'oeil painting or 3-D glasses. In engineering meetings, one of the most withering criticisms of a new feature is that it "goes to 11" — that is, it is more artifice than a useful tool.

Numbers may also not provide a clear guide in evaluating computers. One of the biggest, and sometimes least important, questions facing a computer buyer is how fast a computer should be. The questions may be: "Will a laptop that works at 133 megahertz be fast enough for e-mail and Web surfing?" "Is 350 megahertz enough for a new game?"

Megahertz are usually sold to consumers like the numbers on an amplifier. A higher number is thought to mean greater speed. But people often pay more for performance they will never get.

There is no simple explanation of what a megahertz number means for performance. Generally, 600-megahertz machines run faster than 400-megahertz machines, but they rarely produce the 50 percent increase in performance the numbers would indicate.

Hertz is the unit used to measure the frequency of electromagnetic radiation, which includes things like radio waves. It is used to measure the speed of computers because the machines depend on another form of electromagnetic radiation: electricity. The guts of a computer, the central processing unit, containing a chip, coordinates its work with a central clock switch that also oscillates millions of times per second. That is, a 300-megahertz Pentium processor has a switch inside it that completes a cycle of turning on and off 300 million times per second, and this switch controls the signal that synchronizes the rest of the chip.

Imagine that the central processing unit acts like an assembly line for producing numbers, much as a car factory produces cars. The pieces of data, in the form of pulses of electric current, are gathered, coordinated and fed into circuits that do basic arithmetic like addition and subtraction.

The clock signal acts like a metronome for an orchestra as it synchronizes the assembly of data. There are usually several stations in the assembly line for data, and each station must finish its task before the clock signal changes again. When it does, the data advance to the next station.

It would seem, then, that the speed of a computer is controlled by the speed of the clock signal, so a 300-megahertz machine would appear to be 50 percent faster than a 200-megahertz machine.

The problem with measuring a computer's speed according to the speed of the central processing unit's clock emerges when the data assembly line runs out of data and must pause. The central processing unit is kept waiting when it must fetch data from some distant location, like the memory chips or the disk drive.

When a computer user playing a game goes into a new virtual room, the central processing unit may have to get the information on how many acid-spewing monsters are lurking. Apart from how fast a disk drive is, pieces of data may move slowly because they must travel inches instead of less than a tenth of an inch, and the wires on the circuit boards are often not built to move the current that carries information as fast as the central processing unit does. These wires are usually referred to as a systems bus, a term that borrows the word "bus" to describe how the data are moved in large groups.

The speed of the systems bus, also measured in megahertz, is often a better measure of performance than the speed of the central processing unit. The issue is how many pieces of data are being moved. Central processing units contain a small amount of fast memory, called cache memory, that can keep up with a 300-megahertz clock. But a lot of monsters, or a complicated math problem, would involve too much data to fit in cache memory. In this case, a 300-megahertz chip connected to a 50-megahertz system bus will often sit idling for 250 million of its 300 million cycles each second.

Details, Details: Things Fall Apart

A computer should operate like a well-run hotel. But anyone who owns a computer — like anyone who travels — knows that even meticulously maintained systems break down for reasons that can seem mysterious from the outside.

Within a computer a tiny piece of code may have been omitted when software was upgraded. In the hotel, some crucial employee may be on vacation. In either system there will be gaps and breakdowns that may be difficult to pinpoint and correct.

For instance, in addition to offering directions and recommending area restaurants, an experienced concierge knows the special needs of regular guests. One guest may often receive roses, perhaps, and the concierge may know that the guest is allergic to other kinds of flowers. So the concierge makes sure that if the guest is sent daisies, they are not brought up to the room.

If programmers were writing the code for a digital concierge, they could easily overlook this special duty. Here is how things go wrong in software and, by analogy, in hotels.

❶ TEMPORARY DOORMAN
In the hotel the doorman should not accept any deliveries of daisies for Mrs. Jones. He is supposed to pass them on to the concierge, who knows that Mrs. Jones is allergic to them and accepts only roses. The temporary doorman does not know this.

🖥 FIREWALL
A computer firewall is supposed to keep out a barrage of unwanted e-mail, but sometimes new software can cause holes in the firewall that let through such messages, known as spam.

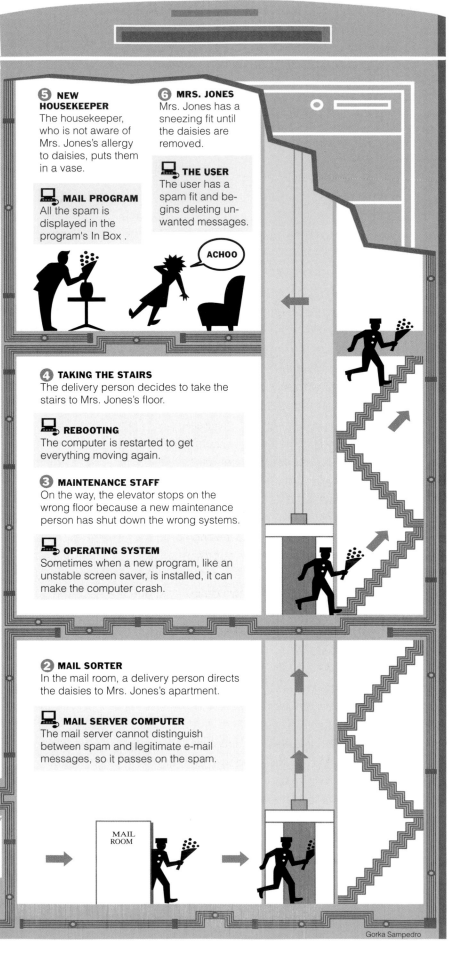

❺ NEW HOUSEKEEPER
The housekeeper, who is not aware of Mrs. Jones's allergy to daisies, puts them in a vase.

🖥 MAIL PROGRAM
All the spam is displayed in the program's In Box.

❻ MRS. JONES
Mrs. Jones has a sneezing fit until the daisies are removed.

🖥 THE USER
The user has a spam fit and begins deleting unwanted messages.

ACHOO

❹ TAKING THE STAIRS
The delivery person decides to take the stairs to Mrs. Jones's floor.

🖥 REBOOTING
The computer is restarted to get everything moving again.

❸ MAINTENANCE STAFF
On the way, the elevator stops on the wrong floor because a new maintenance person has shut down the wrong systems.

🖥 OPERATING SYSTEM
Sometimes when a new program, like an unstable screen saver, is installed, it can make the computer crash.

❷ MAIL SORTER
In the mail room, a delivery person directs the daisies to Mrs. Jones's apartment.

🖥 MAIL SERVER COMPUTER
The mail server cannot distinguish between spam and legitimate e-mail messages, so it passes on the spam.

CIRCUITS HOTEL

MAIL ROOM

Gorka Sampedro

Adding Software Can Add Trouble

By Peter Wayner

Programmers are often good wordsmiths, and one of the better terms floating around today is "bit rot." what happens when software that once worked well fails unexpectedly. The problem often leads many people to avoid expanding their systems or installing new software because they are afraid that old but important programs will stop working.

The term is not literally accurate. The bits that make up a program do not actually change or decompose; they stay exactly the same. The changes crop up in the operating system or in supporting software that does not behave the same way when a new version emerges. An old program may count upon some behavior from an operating system and find that the new version of the system does not deliver it.

Here is a simple example of how bit rot can occur. Imagine that a program uses an electronic dictionary to check grammar in a text. The program is fairly simple, however, and does not anticipate the modern habit of making some nouns into verbs. The program works well and approves sentences like "Their architect flew to France" until a colloquial version of the dictionary is installed that labels words like "architect" as both a noun and a verb. Suddenly, the program starts marking the sentence as having two verbs and no subject.

In most cases, the bit rot is not so easy to understand or diagnose. Programs will often simply crash or start producing gibberish on the screen. As software gets more complex and interconnected, the problem of bit rot potentially becomes worse.

One common problem involves the structure of operating systems like Windows. Most of the Windows services, for example, are broken up into smaller parts known as dynamically linked libraries, or D.L.L.'s. One D.L.L. may contain the software that will draw a window on a screen and fill it with icons representing the files. Another may control access to the Internet. When other software is installed on the computer, it may come with different versions of these basic D.L.L.'s that may then cause incompatibility problems.

For instance, when Microsoft developed a new version of its Internet Explorer browsing software, it wanted to use the same tools for displaying Web pages that it used for drawing windows and icons. So when the new browser software was installed, it replaced an old Windows D.L.L. with a newer version that would also handle Web data. This change and others like it caused some older programs to stop working on many computers.

When they add features, programmers try to maintain the older ones as well, but they often fail to anticipate all the details. Old programs that counted on the old D.L.L. to handle the icons may crash if the new D.L.L. does not handle the process in exactly the same way.

The problem is not limited to D.L.L.s or Microsoft operating systems. Apple has always done a good job with the Mac operating system, but it has not avoided the problem completely.

And many software developers who use the Java programming language silently curse all the browser creators because each browser version creates new incompatibilities that crash their small Java programs, called applets.

To a large extent, users with a blind addiction to new features bring bit rot upon themselves. To appeal to them, software companies concentrate on adding new features instead of removing old glitches. The companies know that new features pry money from the users' wallets twice: Buyers are enticed to grab the latest software version, and then they have to upgrade their older software when it does not work anymore.

To build in an incentive for perfecting existing software, instead of just focusing on new features, some companies have experimented with software licenses and subscriptions that bring in a steadier flow of income. That provides some income to support work in fixing incompatibilities and reducing bit rot. In the long run, bit rot may ebb only after users start recognizing stability and fidelity as important features.

Just Passing Through

In a liquid crystal display, crystals are sandwiched between two pieces of glass and subjected to electrical fields that affect the way light passes through.

L.C.D. MOLECULES GET IN LINE

Although liquid crystals lie in a loose fashion in their natural state, they line up when they come into contact with a finely grooved surface. The alignment layer of an L.C.D. is designed to position the molecules so that they lie parallel to one another along the grooves.

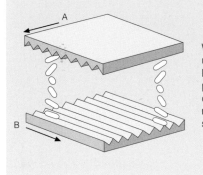

When sandwiched between upper and lower alignment layers that have grooves pointing in different directions, L.C.D. molecules move naturally into a twisted spiral structure.

L.C.D. STRUCTURE

A combination of polarizing filters, transistors, electrodes and liquid crystals work together to control light.

PIXEL IS LIGHT
When two polarizing filters are arranged along perpendicular axes, the light travels along the spiral arrangement of the liquid crystal molecules. That changes the light's orientation so it passes through the second filter and is seen.

PIXEL IS DARK
When an electrical charge is applied, there is no spiral structure to reorient the light, so the light is blocked by the second filter.

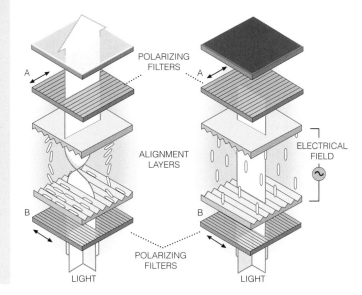

In a Liquid Crystal Display, Molecules Keep Light in Line

By Catherine Greenman

Considering the size of a typical color television, you might wonder how you can also watch your favorite shows on a flat-screen model small enough to fit into a pocket. And if you're considering trading in that bulky computer monitor for a sleek flat-panel display, you might wonder how something that used to take up such an enormous amount of space can squeeze into a case only a couple of inches deep.

Such incredible shrinking acts are possible because of liquid crystal displays, which were first introduced in calculators in the early 1970's and now are used in everything from cellular phones to handheld televisions and desktop and laptop computers.

Traditional cathode-ray-tube television and computer monitors have vacuum tubes that contain electron guns, which shoot electron beams toward a screen covered with phosphors, materials that glow when electrons hit them.

L.C.D. modules apply electrical charges to liquid crystals, rod-shaped molecules that have both liquid and solid properties. The structure of the liquid crystals changes in response to electrical charge, changing the way light passes through them.

In most L.C.D. devices, the layer of liquid crystals, about 5 microns thick (a human hair is about 100 microns wide), is between two layers of plastic or glass, both of which are covered with a layer of transparent electrodes, and two alignment layers; each alignment layer orients the liquid crystal molecules in a fixed direction. Light passes through a polarizing filter, then the L.C.D. cell, then another polarizing filter. The two filters are oriented in such a way that only light that has changed its orientation while passing through the liquid crystals will pass through the top filter and be seen.

There are two types of L.C.D. devices, active matrix and

① Light for L.C.D. monitors is produced by a set of fluorescent bulbs either along the edges of or behind the glass panel at the back of the display.

② The light passes through one polarizing filter before passing through the liquid crystal; only light that has had its polarization rotated 90 degrees by twisted, spiral liquid-crystal structures can pass through a second polarizing filter and get to the viewer's side of the glass. How much light makes it through depends on how much of the liquid crystal layer is twisted into a spiral.

③ In an active-matrix display, each liquid-crystal cell has at least one transistor, which turns the cell on or off like a switch. In color active-matrix displays, the light passes through a color filter before reaching the second polarizing filter. Each pixel has three fixed transistors, one for every red, green and blue cell.

④ The greater the electrical field, the more the liquid-crystal molecules move out of their spiral structure. That means that less light is rotated and less of it gets through the second polarizing filter.

⑤ In this example, the liquid crystals in the cell sending light to the blue filter are not in a spiral structure, so no blue light can get through the second polarizing filter. And only some of the liquid-crystal layer is in a spiral in the cell sending light to the green filter, so only some green light gets through. But all of the liquid-crystal layer is twisted into a spiral in the cell sending light to the red filter, so all the red light makes it to the viewer's side of the screen.

COLOR ACTIVE-MATRIX DISPLAY

Unlike passive-matrix displays, which use relatively few transistors arranged in rows along the liquid-crystal layer, active-matrix displays (like the one shown here) have an individual thin-film transistor for each of the three cells in a pixel. The transistors help drive the activity of the cells as they pass and block light, resulting in a brighter, sharper display than passive-matrix screens.

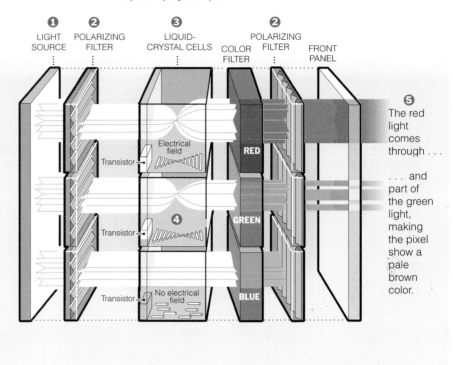

Gorka Sampedro and Mika Gröndahl

passive matrix. Active-matrix L.C.D.'s are used in products like more expensive laptop computers, flat-panel monitors and L.C.D. projectors. The primary difference is that an active-matrix display has at least one transistor for each of the pixels, the elements that make up the display (50,000 to several million, depending on the resolution), and that gives a finer control over the light exiting the L.C.D. cell. A black-and-white active-matrix display has one transistor per pixel. Each color active-matrix L.C.D. pixel has three cells — red, green and blue — with one transistor per cell. Because of the fine control made possible by the transistors, the liquid crystals, which block or pass different kinds of light, produce thousands of different color variations. If you notice an aberrant red, green or blue spot (also known as a dead pixel) on an active-matrix display, it means that a transistor is not functioning properly.

Passive-matrix L.C.D.'s, which are used to display information on calculators, cellular phones, handheld computers and many other electronic devices, do not have transistors at every pixel. Instead, a charge must be sent all the way across the display at the intersection of each pixel. Sending charges all the way across or all the way down can inadver-

tently affect other liquid crystals along the way. That is why you will sometimes see "ghosting," or blurred images, on a passive-matrix L.C.D. screen.

Because each pixel in a row or column in a passive-matrix display is charged in succession, much the way a freight train rolls down a track, the passive display has a generally lower response time than an active-matrix display. As a result, their images are not as bright, video can appear jumpier, and viewing angles are limited — you have to look directly at the display to see what's on the screen.

Most L.C.D. monitors and computer screens are lighted by small fluorescent bulbs, sometimes no larger than the circumference of a straw, from the rear of the display. Displays with backlighting also use a reflector to shine the light source through the display to utilize as much of the light produced as possible.

L.C.D.'s in devices like calculators have no backlights and use reflected light, so they work only in a bright room. Light shines through the front polarizer and glass and through the crystals and rear polarizer. It then bounces back through the rear polarizer and crystals toward the viewer.

A Sandwich of Layers

Touch screens have to perform two separate tasks: they must display a screen full of items to point at, then must detect which ones you're pointing at. The display screens are like television sets or computer monitors, but there are four different technologies for detecting the touch of a finger or stylus.

Resistive overlay screens

Used for point-of-sale terminals in stores, hotels and restaurants, and, as shown here, in hand-held computers like the Visor.

A resistive screen is a sandwich, with two conductive layers that are kept apart, usually by a grid of tiny insulating dots. Pressure from a finger or stylus pushes the two layers together, creating an electrical contact.

.003 in.

STYLUS

CONDUCTIVE LAYERS

SCRATCH-RESISTANT COATING

POLYESTER LAYER

SPACER

CONTACT POINT

GLASS

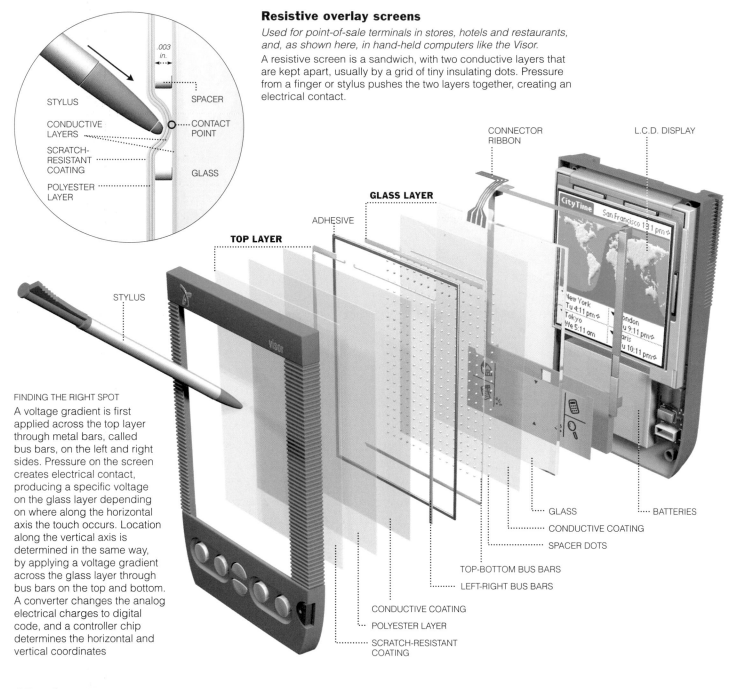

CONNECTOR RIBBON

L.C.D. DISPLAY

GLASS LAYER

ADHESIVE

TOP LAYER

STYLUS

CityTime   San Francisco 1:11 pm

New York
Tu 4:11 pm
Tokyo
We 5:11 am

London
Tu 9:11 pm
Paris
Tu 10:11 pm

GLASS

BATTERIES

CONDUCTIVE COATING

SPACER DOTS

TOP-BOTTOM BUS BARS

LEFT-RIGHT BUS BARS

CONDUCTIVE COATING

POLYESTER LAYER

SCRATCH-RESISTANT COATING

FINDING THE RIGHT SPOT
A voltage gradient is first applied across the top layer through metal bars, called bus bars, on the left and right sides. Pressure on the screen creates electrical contact, producing a specific voltage on the glass layer depending on where along the horizontal axis the touch occurs. Location along the vertical axis is determined in the same way, by applying a voltage gradient across the glass layer through bus bars on the top and bottom. A converter changes the analog electrical charges to digital code, and a controller chip determines the horizontal and vertical coordinates

Other types

CAPACITIVE OVERLAY
Kiosks, gambling machines.
A glass overlay is coated on both sides with a transparent layer of metal, with a protective glass layer on top. Current is passed through the metal coating. When a finger touches the screen, minute electrical charges on the finger attract current from the screen. Electrical flow is measured from each corner to determine finger position.

SCANNING INFRARED
Automatic tellers, medical instruments, kiosks, vehicles.
The screen is surrounded by rows of light-emitting diodes and detectors. The L.E.D.'s send infrared beams across the screen. A finger blocks some of the beams. A controller chip calculates the finger position from "no light" signals from the detectors.

SURFACE ACOUSTIC WAVE
A.T.M.s, subway tickets, gambling and kiosks.
Similar to infrared, but uses ultrasonic waves produced by piezoelectric transducers. Transducers on the opposite side of the screen convert the ultrasound back into electrical signals. The finger absorbs some of the ultrasound energy, changing the signal, which the controller uses to map the point touched.

Sources: Handspring; MicroTouch Systems Inc.

Frank O'Connell

With Touch Screens, It's O.K. to Point

By Matt Lake

Sometimes a computer mouse just isn't easy enough to use. When bank customers are taking money from automated teller machines or restaurant workers are tallying bills, grabbing an electronic device encased in a handful of plastic doesn't come naturally at all.

That's when people find the original pointing device — the index finger — much more efficient and easier to use. That is why touch-screen technology was developed and why it is now used for everything from palm-size computers to controllers in airplane cockpits.

Touch screens combine two separate technologies, one to display items on a screen and the other to figure out what you are pointing at. The displays are either cathode-ray tubes, like those used in televisions, or flat panels, like those in laptops.

Every part of a touch-screen system must work fast. Without a visible change or audible cue, most people touch the screen repeatedly (which can overload the system) or assume that the device is broken.

Touch screens use one of four different technologies to detect touch, and electronic controllers and software drivers to communicate that information to a computer. The touch sensor system must be calibrated so a touch can be correlated exactly to the right spot. Two of the technologies, called resistive overlay and capacitive overlay, can drift out of synchronization and need to be recalibrated periodically. The other two use ultrasound or infrared signals and tend to stay aligned.

The earliest touch-screen technology, which is also the most widespread and least expensive, is the resistive overlay. Resistive overlays were first developed by Elographics, a company in Oak Ridge, Tenn. Elographics (which has since changed its name to Elo TouchSystems) came up with the idea of overlaying two conductive sheets separated by a mesh of tiny raised dots, then passing electricity through one layer. When the screen was touched, it would make contact between the two layers, letting the electricity take a shortcut.

That interruption of the flow of electricity is at a measurable point — one that can be determined precisely enough to register the touch of a stylus tip.

The resistive overlay system has a few drawbacks, but it is the system used in hand-held computers like Palm organizers. The overlay is susceptible to abrasion, which rules out its use in public kiosks. And the thin conductive layers do block some light from the display screen.

More light makes its way through capacitive overlay touch screens, which helped that technology catch on, especially for gambling machines, where graphics are important. When the screen is touched, the user draws a small amount of electricity to the point of contact, so that touch can be detected. The capacitive coating is thinner and harder than the glass it is applied to, which makes it more durable for public uses. But the user cannot wear gloves or point at the screen with a stylus or pen.

That gave rise to another type of touch sensor system, called surface acoustic wave technology, that combines durability with sensitivity. Surface wave touch screens do not require any opaque overlay so their displays are bright — and they do not require periodic recalibration. These screens detect the absorption of ultrasonic waves at the point of touch.

Like capacitive screens, acoustic wave screens are robust. That's why they are used in machines dispensing Metrocards for riding the subways and buses in New York City. Many banks use the technology in automated teller machines. The screen can become quite dirty and dusty without giving false readings.

The most recent addition to the touch-screen pantheon, infrared screens, use a grid of infrared light and sensors to detect the point of touch. Infrared screens tolerate a range of temperatures and do not require frequent recalibration, which makes them suitable for use in airplane cockpits and A.T.M.'s alike.

Painting an Image in the Eye

A wearable retinal scanning display system by Microvision uses a laser and a tiny scanner to sweep an image pixel-by-pixel across the back of the eye. The scanner, a MEMS (short for microelectromechanical systems) device, is manufactured using standard semiconductor fabrication techniques.

1 A red laser diode emits a pulsed beam containing the image in a pixel-by-pixel stream. Electronics monitor the beam's strength to ensure it will not harm the eye.

2 The beam hits the mirror on the MEMS scanner, which reflects it toward the optical combiner. The mirror moves in two directions, creating the scanning pattern.

3 Lenses in the optical combiner divert the beam through the pupil into the eye.

4 The beam sweeps across the retina in a horizontal line, hitting photoreceptors one pixel at a time. When it reaches the end of each line, the beam moves back to the other side and down slightly to draw the next.

The eye combines the laser image with the real-world image. In the example below, the laser provides grading information and contours for a heavy-equipment operator.

OPTICAL COMBINER

DISPLAY MODULE
(head-mounted)

RED LASER DIODE

MIRROR

MEMS SCANNER

CONTROLLER MODULE
(belt mounted)

From computer

The headset

The display module and see-through optical combiner are mounted on a headset and connected by cable to the controller module, which is worn on the belt. A complete wearable system would also include a small belt-mounted computer.

Key components

CONTROLLER
The controller takes the video signal from a computer and converts it into two streams of data: one controls the laser beam, and the other controls the scanner.

LIGHT SOURCE
The light in a retinal display comes from a laser diode that produces a beam that is sufficiently weak that it cannot damage the retina (something industrial-strength lasers can do). Full-color displays use three light sources, in red, blue and green.

MEMS SCANNER
The scanner's mirror is about a tenth of an inch wide. An electromagnetic system causes it to pivot independently along two axes, creating the scanning pattern. All of this happens very quickly because the laser beam transmits millions of pixels per second. Although only one pixel hits the eye at any given fraction of a second, the stream is fast enough to trick the eye into perceiving a full picture.

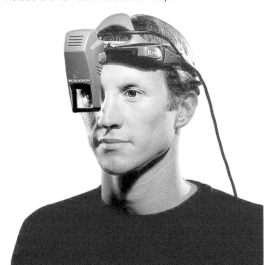

Source: Microvision, Inc.

Frank O'Connell

Double Vision: Retinal Displays Add a Data Layer

By Matt Lake

There is a scene in "The Terminator," the 1984 James Cameron movie, in which the audience sees through the robotic eyes of the title character, played by Arnold Schwarzenegger. When the superintendent of a building yells at the Terminator, a list of possible responses scrolls down the screen before he chooses one to shout back. The one he picks is a doozy.

In the 1980s, science fiction movies were the only place you could see a computer-generated image on top of a real-life background. Now, a commercial product achieves the same effect in real life by projecting images directly onto the viewer's retina, at the back of the eye.

Head-mounted displays are not new. Several companies market L.C.D.-based displays for wearable computers. But the Nomad system, developed by Microvision, Inc., of Bothell, Wash., is different: By projecting images onto the retina, it superimposes data on what is viewed without hampering the person's vision.

On the surface, wearable scanning displays have more in common with bicycle helmets than with the computer monitors and televisions they may replace. The Nomad, for example, is mounted at the front of a molded plastic headpiece; its box of optical, electrical and mechanical components has a single window that can slide over either eye. The components are linked to a computer that can be worn on a belt.

The Microvision displays do not use a screen, although they work somewhat like a television set. Just as a cathode-ray tube in a television or monitor scans a beam of electrons onto the front face of the tube, the tiny projector inside a head-mounted display beams light through the pupil and onto the nerve cells in the retina.

At first glance, pointing a laser directly into your eyeball seems to fall somewhere between the risky and the downright foolhardy. But Microvision's retinal scanning devices comply with the safety rules set by the American National Standards Association and the International Electrotechnical Committee. According to Microvision representatives, the most recent standards for light levels allow beams that are almost 100 times as bright as those emitted by its retinal scanning displays.

Because the technology uses a scanning beam to paint the image on the retina, Microvision calls a device using this technology a retinal scanning display — not to be confused with the biometric technology of retinal scanning, which detects the distinctive pattern of blood vessels on the retina, a pattern that can be analyzed to identify someone.

Although the display system is proprietary, it is based around open standards. It accepts television signals in the American NTSC and European PAL formats, as well as in the SVGA graphics format, which can display an image 800 pixels (or dots) wide by 600 pixels high.

The first wave of applications for scanning displays is geared for military, medical and aerospace users, in part because the cost is expected to be around $10,000 for a single unit. The Nomad system, could be used in a hospital, for example, to superimpose information about a patient, including vital signs and medical images, onto a surgeon's field of view.

The display could also be used for the controls of industrial machines and aviation, including air traffic control. In industrial settings, for example, it could give service technicians access to manuals. Drivers of heavy equipment would also be able to benefit from displays of maps and instrument readings as they keep their eyes on the road ahead.

A heads-up display can also be used in dangerous situations, giving firefighters in a smoke-filled building a floor plan, for example, or warning workers of gas leaks.

But the devices may someday be marketed for consumers. Eventually, it may be possible to relax on the couch to watch television with retinal scanning technology. So instead of experiencing James Cameron's dystopic vision of killer robots from a safe distance on a video screen, you could end up watching his movies through their eyes instead.

Follow the Spinning Ball

Computer mice are of two general types, optical and mechanical. The heart of a mechanical mouse is a ball that can move in any direction when the mouse is rolled over a pad or other uniform surface. The forward-and-back and side-to-side motion of the mouse is converted to up-and-down and side-to-side movement on the computer screen.

BUTTONS

These are actually small switches. Pressing the button closes a circuit, which is interpreted by the computer as a command.

BALL

The ball rests on three shafts, one for support and two to gather directional information. When the ball moves, one shaft turns in response to forward-and-back motion, while the other tracks side-to-side motion. Each of those shafts contains a slotted wheel that also rotates. Better-quality balls have a metal core (partly shown) for weight and stability.

INTERFACE INTEGRATED CIRCUIT

A chip takes information on movement and information about whether a switch is activated and combines that into a uniform data stream. The information is sent to the computer, where it is processed. A signal from an activated switch is interpreted as a command, while motion is translated into a change in cursor position.

Sources: Microsoft; Logitech

SWITCH

X-AXIS SHAFT

SUPPORT WHEEL

Y-AXIS SHAFT

DRIVERS

Software drivers interpret the direction and speed of the mouse movement and produce a corresponding movement of the cursor.

Behind the Lowly Mouse: Clever Technology Close at Hand

By Howard Alexander

While many computer users think that the mouse was born when Apple introduced the Macintosh computer in 1984, the mouse is actually two decades older than that.

What came to be called the mouse was developed in 1964 by Douglas Engelbart, a professor working at the Stanford Research Institute in Menlo Park, Calif. The mouse was invented as a way to make it easier to move the cursor around the screen of early computers, and it proved to be a more efficient pointing device than the arrow keys of a keyboard. One person in the laboratory thought that the unit looked something like a mouse, and the name stuck.

The first mouse was a mechanical unit shaped like a small brick, with one button on top and two wheels underneath for detecting vertical and horizontal motion. It was somewhat

awkward to maneuver, but when the National Aeronautics and Space Administration tested the mouse in 1966 against several other devices, like a light pen and a knee switch, the mouse was found to be the superior technology (although NASA scientists worried that a mouse might float away from the work surface in the weightless environment of space).

Flashing forward to the next decade, we arrive at Xerox's Palo Alto Research Center. Here we find Dr. Engelbart's mouse being combined with the first graphical user interface in a small computer. The Alto, introduced in 1973, one of the first personal computers on the market for general use, inspired both the Macintosh and Windows operating systems. Xerox sold only a few Altos and ultimately withdrew them from the market. At that point, the wheels of the mouse had

Tracking the Movement

As each slotted disk rotates, the light shines through the slots and is blocked by each tooth of the disk, creating a flickering pattern of light and dark. The pattern is converted by the phototransistors into a series of pulses that are sent to the interface integrated circuit. Moving the mouse faster causes the wheel to spin faster, making for shorter pulses that the software driver translates into faster cursor movement on the screen.

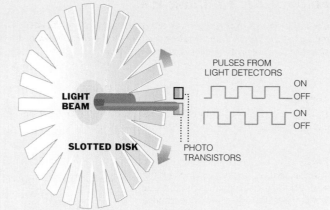

PULSES FROM LIGHT DETECTORS

LIGHT BEAM

SLOTTED DISK

PHOTO TRANSISTORS

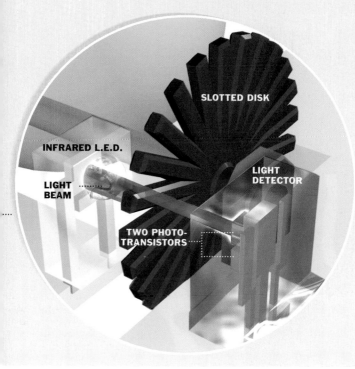

SLOTTED DISK

INFRARED L.E.D.

LIGHT DETECTOR

LIGHT BEAM

TWO PHOTO-TRANSISTORS

SENSING THE DIRECTION OF THE MOUSE

Each disk assembly has two phototransistors next to each other. The light pulses created by the spinning disk fall on the two phototransistors at slightly different times. The patterns are the same, but how they are out of sync indicates which direction the mouse is moving — from left to right or right to left, for example.

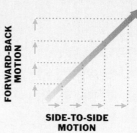

FORWARD-BACK MOTION

SIDE-TO-SIDE MOTION

COMBINED MOTION

In the computer, the information about the components of the mouse's motion is combined and translated into cursor motion.

Mika Gröndahl

been replaced by a single ball, like that in the modern mouse, and the unit had three buttons. It was called a pointing device as well as a mouse.

The device proved useful for many purposes. When reasonably priced computer systems became available for businesses of all sizes to do computer-aided design, the mouse proved to be an ideal device for drawing and drafting. These systems enabled the user to make designs and changes rapidly and efficiently. Air traffic control systems embraced the mouse and its upside-down sibling, the trackball, which was developed in the 1950s as part of display and tracking systems. Bars and arcades found another use for trackball technology: video games.

With the growth of home computers, the mouse slowly started to make itself at home on the desk. Users found it a convenient way to move the cursor around the screen. Sales of mice went up, and prices started to go down. Mouse Systems patented an optical-electronic sensor system that eliminated the mechanical sensors in the mouse and made them more reliable.

When Apple made a big splash with the Macintosh in 1984, the Mac operating system brought the mouse to every-

one's attention. Other operating systems with graphical user interfaces (like Commodore's Amiga, Digital Research's GEM, IBM's Topview, Microsoft's Windows 1.0, and Visicorp's Vision), some even preceding the Mac, generated a lot of interest, at least in the press, and all of them used mice.

Since then, the mouse has continued to change. Mouse System's PC Mouse, introduced in 1986, uses lights bouncing off a reflective mouse pad to detect motion and has no moving parts. Other manufacturers, like Logitech, have come out with wireless mice, and some vendors have tried different shapes.

Both the mouse and the trackball continue to propagate and evolve, well, like mice. In addition to the different shapes, there has been a trend toward adding more functions and greater reliability. Some new mice, for example, have a wheel or button that can be used for scrolling.

A mouse does require some attention. Most mice and trackballs are based on the principle of ball rotation. With a mouse, the ball picks up dust from the desktop and transfers the dirt to the internal mechanism. That can make the cursor appear to jump or skip. A clean mouse, therefore, starts with a clean desk.

Cursor Control With a Twitch

Touch pads were introduced in 1994 as an alternative input device for laptop computers, but their accuracy was limited. Now that touch pads have been improved, they are also available as optional pointing devices for desktop computers.

Electrode layers maintain a constant AC signal when the pad is untouched, and the signal's status is monitored by the circuit board. When a suitably conductive object, like a finger, approaches the pad, the signal is distorted. The horizontally and vertically aligned electrode bands create a grid from which the specific point of contact is determined.

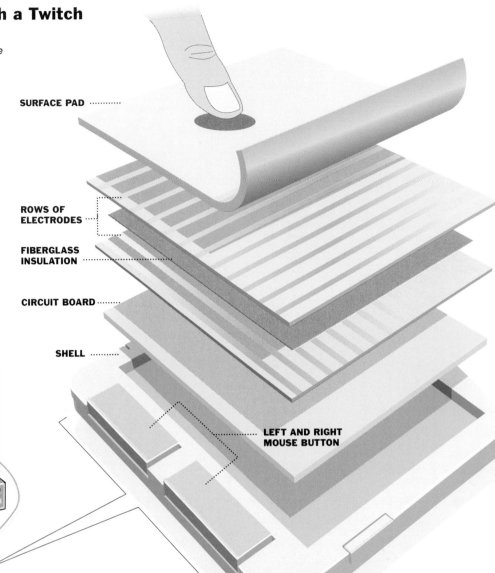

SURFACE PAD

ROWS OF ELECTRODES

FIBERGLASS INSULATION

CIRCUIT BOARD

SHELL

LEFT AND RIGHT MOUSE BUTTON

Fingertip Precision

❶ ELECTRIC-FIELD CHANGES

At many points along the touch pad, an upper and lower electrode are separated by a thin layer of fiberglass insulation. When a voltage is applied, each of those points becomes a capacitor, which stores charge and has an electric field. When a finger is near, it distorts the electric field.

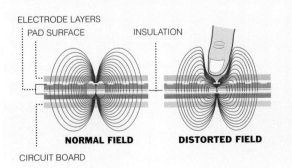

ELECTRODE LAYERS
PAD SURFACE   INSULATION

NORMAL FIELD   **DISTORTED FIELD**

CIRCUIT BOARD

Sources: Cirque Corporation

❷ TRACKING FINGER MOVEMENTS

Looking at each pair of electrodes to find out where the electric field is disrupted would be too slow. So patterns of pulses of electric current are run horizontally and vertically, and any changes caused by a finger are measured by the circuit board. By combining the vertical and horizontal information, the circuit board quickly tracks the finger's position on the touch pad.

Each time the position of the finger changes, the logic board sends this information to the user's computer and the cursor responds accordingly.

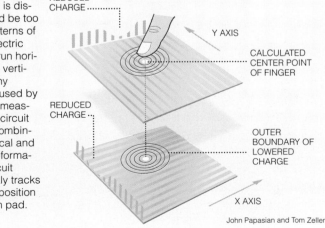

REDUCED CHARGE

REDUCED CHARGE

Y AXIS

CALCULATED CENTER POINT OF FINGER

OUTER BOUNDARY OF LOWERED CHARGE

X AXIS

John Papasian and Tom Zeller

Treading on the Mouse's Heels: The Oh-So-Subtle Touch Pad

By James Ryan

For many people, the computer mouse is a marvel of ingenuity. But for George E. Gerpheide, an engineer, it is an abomination. He set out to build, if not a better mousetrap, then something to rid society of the inelegant mouse. "It seemed clunky," said Dr. Gerpheide. "What the world needed was a pointing device that matched people's expectations and looked like it belonged in the 21st century."

After two years and 19 unsuccessful prototypes (built after hours in the basement of his home), he hit upon the solution in 1988: an electronic touch pad that would register the slightest twitch of a finger and translate it into cursor movements on the monitor. He picked the nickname "cat" for his rival to the mouse and founded the Cirque Corporation to market the technology.

Apple was the first to license Dr. Gerpheide's technology and take it to market. It incorporated a touch pad 1 1/2 inches wide into its Powerbook laptops in 1994, replacing the built-in trackball. Most laptop manufacturers, with the exception of I.B.M. and Toshiba, which employ a joystick, have adopted touch pads since then.

Touch pads range from the size of a pack of matches to that of a small wallet. They rely on minute flows of electric current and how those currents are modified by the human finger's effect on small electric fields, so moisture or electromagnetic interference can diminish performance. That was particularly true of early touch pads.

What touch-pad users do not have to worry about is peanut butter, cookie crumbs or hair fouling any moving parts, as they can in a mechanical mouse. They also do not need to make space on their desks for a mouse to roam.

Unlike touch screens and some touch pads called resistive, which require pressure from a finger or stylus to function, one need not actually touch the surface of a capacitive touch pad to make it work. The finger can affect electric fields in the touch pad even when it is merely near the surface. But if a business card or two is placed on top of a touch pad, that extra insulation will keep it from registering the presence of a finger running on the surface.

Another advantage of newer models of touch pads is that they can be trained to recognize handwritten commands. Synaptics sells a model, popular in China, that enables users to use their fingertips to enter Chinese characters into the computer.

If used properly, the touch pad is less likely to cause repetitive stress injuries because touch-pad users do not have to grip or press anything. The finger movements required are subtle and relatively easy.

Dr. Gerpheide offered a few tips for using a touch pad: (a) The lighter the touch, the better the touch pad works. (b) For a large cursor movement, slide the finger; for a more precise movement, hold the finger still and roll the tip from side to side. (c) A double tap equals a double click of the mouse. To highlight and drag an object, use a tap and a half: tap once, then lower the finger and leave it down.

Despite the advantages of touch pads, they have been slow to catch on, especially among desktop users, who may have had bad experiences with the first generation of touch pads, which were less precise. There have been many innovations since then, but it takes roughly two years for the product cycle to catch up to new touch-pad technology, he added.

Other users are simply more comfortable with the more familiar mouse, but Dr. Gerpheide is banking that the rapid evolution of technology will change that. "In the post-PC era, when you're talking about mobile appliances," he said, "the mouse is not suitable. Nobody wants a mouse dangling from their Net-connected cell phone."

Translating Data for the Telephone

To send e-mail or other information from one computer to another, binary data (encoded as ones and zeros) is transmitted from the computer to a modem, which changes the data into analog information — sound waves — that can be sent over a telephone line. The modem at the receiving end reverses the process, changing the analog data back into binary code that the computer can understand.

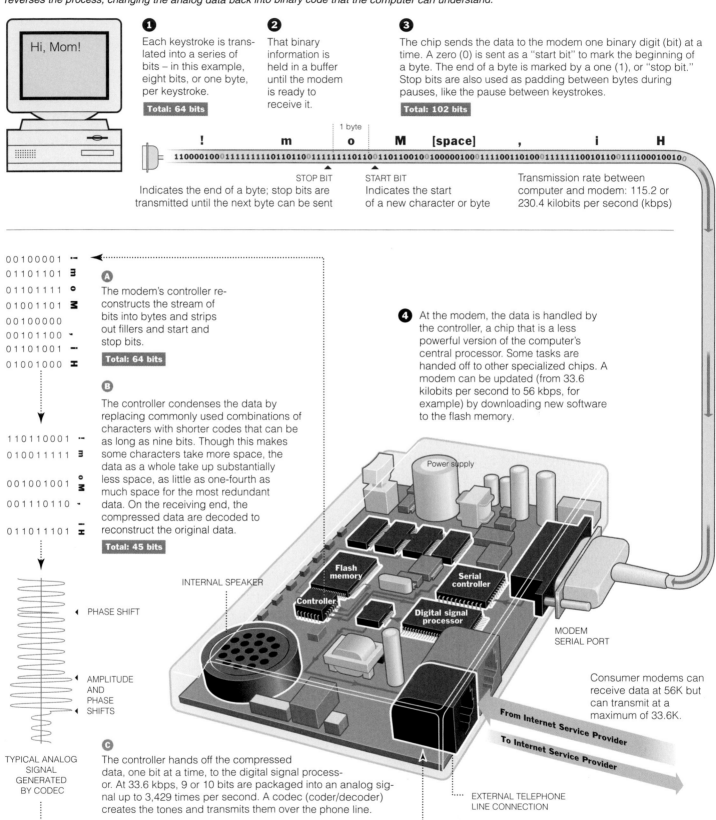

Hi, Mom!

❶ Each keystroke is translated into a series of bits – in this example, eight bits, or one byte, per keystroke.

Total: 64 bits

❷ That binary information is held in a buffer until the modem is ready to receive it.

❸ The chip sends the data to the modem one binary digit (bit) at a time. A zero (0) is sent as a "start bit" to mark the beginning of a byte. The end of a byte is marked by a one (1), or "stop bit." Stop bits are also used as padding between bytes during pauses, like the pause between keystrokes.

Total: 102 bits

1 byte

!   m   o   M   [space]   ,   i   H

11000010001111111110110110011111111011001101100100100000100011100110100011111110010110011100010010

STOP BIT
Indicates the end of a byte; stop bits are transmitted until the next byte can be sent

START BIT
Indicates the start of a new character or byte

Transmission rate between computer and modem: 115.2 or 230.4 kilobits per second (kbps)

00100001 ┇
01101101 ┇
01101111 o
01001101 ☰
00100000
00101100 ·
01101001 ┇
01001000 ⌶

Ⓐ The modem's controller reconstructs the stream of bits into bytes and strips out fillers and start and stop bits.

Total: 64 bits

Ⓑ The controller condenses the data by replacing commonly used combinations of characters with shorter codes that can be as long as nine bits. Though this makes some characters take more space, the data as a whole take up substantially less space, as little as one-fourth as much space for the most redundant data. On the receiving end, the compressed data are decoded to reconstruct the original data.

Total: 45 bits

110110001 ┇
010011111 ☰
001001001 o ☰
001110110 ·
011011101 ┇

❹ At the modem, the data is handled by the controller, a chip that is a less powerful version of the computer's central processor. Some tasks are handed off to other specialized chips. A modem can be updated (from 33.6 kilobits per second to 56 kbps, for example) by downloading new software to the flash memory.

Power supply

Flash memory

Controller

Digital signal processor

Serial controller

INTERNAL SPEAKER

MODEM SERIAL PORT

Consumer modems can receive data at 56K but can transmit at a maximum of 33.6K.

From Internet Service Provider

To Internet Service Provider

◄ PHASE SHIFT

◄ AMPLITUDE AND PHASE SHIFTS

TYPICAL ANALOG SIGNAL GENERATED BY CODEC

Ⓒ The controller hands off the compressed data, one bit at a time, to the digital signal processor. At 33.6 kbps, 9 or 10 bits are packaged into an analog signal up to 3,429 times per second. A codec (coder/decoder) creates the tones and transmits them over the phone line.

EXTERNAL TELEPHONE LINE CONNECTION

Sources: Rockwell Semiconductor Systems; Cass R. Lewart, "The Ultimate Modem Handbook"

Baden Copeland

Linking the Computer and the World, the Clever Modem

By Glenn Fleishman

Pick up a phone while your computer is connected to the Internet, and you hear what sounds like the braying of an electronic donkey. Somewhere inside that set of squawks might be a picture of your grandmother, an e-mail to your boss or perhaps the new software slowly downloading to your computer. Bits and bytes have been turned into audio chatter by a technological workhorse: the modem.

The mystery unravels a bit when you consider a modem's full name: MOdulator/DEModulator. A modem takes digital information, in the form of single binary digits (bits), and passes those bits through a series of steps to repackage, or modulate, them into analog signals — sound waves, in this case — so the bits can be passed over a regular phone line. The process is reversed as the modem deciphers, or demodulates, information coming in from the phone line.

Before modems, computers had to be "hard wired" together, even over long distances, to talk to one another. Modems have made communication cheaper and easier, eventually leading to the Internet. Although phone companies' switching equipment turns analog signals into digital ones for transmission over long distances, the weak leak is still the analog hookup between a home or business and the phone company.

A modem translates between the computer's digital world and the phone line's analog audio world. The computer sends out information packets as a series of ones and zeros. Since a telephone carries information in the form of electrical signals representing sound waves, the quality of transmission can vary widely from place to place. One house might have clean lines, while clicks and radio station transmissions might intrude upon the lines of the house next door. The distance from a switching station can also affect clarity.

A phone line carries only the small range of frequencies in which most human conversation takes place: about 300 to 3,300 hertz. The modem works within these limits in creating sound waves to carry data across phone lines.

But modem designers have become increasingly clever at squeezing more bits into that tonal range. Software that runs in the modem analyzes data for repetitive information, like the space character repeated 100 times in a row or the word "the" occurring over and over. Before transmission, these patterns are replaced with symbols or codes. The receiving-end modem takes the codes and reconstructs the original redundant bits, sending them to the receiving computer without any loss of information.

Modems use different tones to represent different bits of data. Each set of tones is called a symbol. Baud measures the number of symbols, or discrete units of information expressed by phase and amplitude modulation, per second. The measurement is named for Jean-Emile Baudot, a Frenchman who developed a binary code in the 19th century.

Modems were first widely used in the late 1960's. They ran at 110 baud, sending one bit per symbol, or 110 bits per second (bps). The 300-baud modems that showed up in the early 70's were an enormous improvement, akin to the leap from 9,600 bps to 28,800, or 28.8 kilobits per second (kbps).

But the breakthrough that led to today's fastest modems was the idea of sending several tones at the same time, creating a mini-symphony of sounds in each symbol. So starting with the 1,200-baud modems that came into general use in the early 1980's, each symbol has been carrying many bits at once, not just a single bit. Phone line quality has also continued to improve, so more tones can be used over a greater frequency range. A 33.6-kbps modem, for instance, transmits about 9.8 bits of data per symbol (data for correcting errors are also sent), at a rate of about 3,429 symbols per second.

The 56K (56 kbps) modems run no faster than 33.6 kbps "upstream," to an Internet service provider, and about 40 to 50 kbps "downstream," with data coming from the I.S.P. to the user's computer. The transmission is all digital on the I.S.P.'s end, allowing it to do a better job encoding information for sending, but the specialized equipment the I.S.P. uses is not available for a user's modem.

Branching Out

To the uninitiated, Web addresses seem like a jumble of letters, numbers and punctuation. But it does not take long to discern order in the chaos. Most addresses begin with http://, which reflects how computers on the Web trade information: a method called hypertext transfer protocol. That may be followed by www and may end in a suffix like com, org, net or gov. In between are names identifying the desired site. Most Web addresses follow this hierarchy so the root servers keeping track of each Web site's location can find it quickly when a user goes looking. Here's how those servers work.

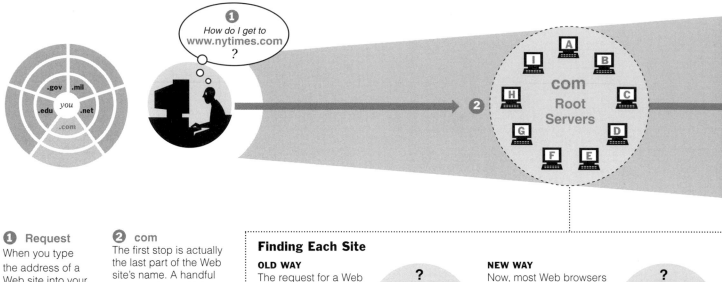

❶ Request
When you type the address of a Web site into your browser, your computer asks other computers, known as servers, for directions.

❷ com
The first stop is actually the last part of the Web site's name. A handful of computers, which are called root servers and are located around the country, keep track of the location of every Web address that ends in com.

Finding Each Site

OLD WAY
The request for a Web page used to go first to the root server called "A" and work its way around to the others until it found the right address.

NEW WAY
Now, most Web browsers remember which of the root servers gave them the response most quickly the last time the question was asked, and they ask that root server first.

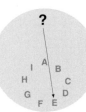

How Servers Find a Needle in the Haystack Called the Internet

By Peter Wayner

The Defense Department created the Internet and so the nature of its offspring should come as no surprise — it's a vast hierarchy with redundancy added in case of war.

That is most evident in the conventions for creating .com Web addresses and the like, which are just aliases for the strings of numbers that computers use for addresses. The addresses are called domain names; they are converted into numbers known as I.P. (for Internet protocol) addresses. I.P. addresses come as sets of four numbers, separated by periods, like 199.183.172.21.

The domain name system, or D.N.S., is the Internet telephone book that allows computers to convert names into numbers. Of all the computers that make up the Internet, the machines that are ultimately responsible for keeping track of all the names are called the root servers.

The pieces of site names fall into a hierarchy. To parse a name in descending hierarchical order, you start at the right.

Americans are most familiar with addresses that end with .com, the designation for commercial ventures. That is called a top-level domain. Other top-level domains are .org for groups like nonprofit organizations, .gov for government agencies, .edu for educational institutions, .mil for military sites and .net for companies that are more closely related to providing Internet services.

Outside the United States, the top-level domains are usually related to a country. For instance, many addresses in the United Kingdom end with .uk. Some addresses in the United States also end in .us, but they are less common.

Top-level domains are just the beginning, even though they come at the end of addresses. Computers decode addresses by examining them from right to left. The machine running the top-level domain answers questions about the smaller domains in the address, and these smaller domains may be broken up even more.

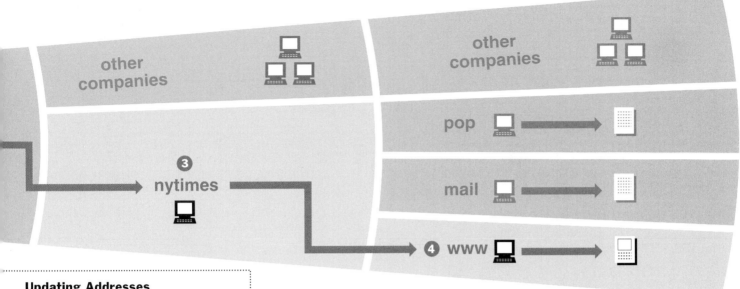

Gorka Sampedro and Kris Goodfellow

Updating Addresses

The "A" root server keeps a master list of every Web address, and each night the master list is copied onto the other root servers.

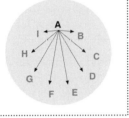

③ The root server directs you to another computer, nytimes, which acts like a telephone operator forwarding requests to yet another server owned by The New York Times.

⑧ The last stop the request makes is www. The www server will bring up the requested Web page. If the request had something to do with e-mail instead, it would have traveled to either the mail or pop servers.

The Web addresses most known to Americans have three levels, but ones with four or even five levels are quite common. Most companies stick their Web servers in the third-level domain and name it with www. But this is not a requirement.

The hierarchy gives computers a clue about how to convert domain names into numbers. Each D.N.S. server is responsible for answering questions about all the machines a step down in the hierarchy. Consider how a computer might look up the www.internetcompany.com address. First, it goes to a root server that maintains the list for the top-level domain at the end of the name, in this case, a .com server. It asks for the address for internetcompany, then goes to that address and asks for the www address.

What happens when the root server suffers a glitch? Is all of the Internet frozen if someone trips over the plug to this machine? No. There are actually numerous root servers for the .com domain and more for others. If a computer cannot find the right answer from one, it asks the others.

It should be immediately clear how valuable it can be to control this first step in looking up an address. One glitch in the database can make someone disappear or send that

person's traffic to someone else. One company, Network Solutions, used to have an exclusive contract from the United States government to both register domain names and maintain the database of most popular top-level domains like .com and .org. The registry business has since been opened up to other companies, but Network Solutions still manages the databases.

For example, Network Solutions runs the primary top-level server for the .com domain. Each day, that primary server takes its master list of names and I.P. addresses and converts it into a database that is then copied to the other root servers.

In 2000, the agency that controls the Internet address system, the Internet Corporation for Assigned Names and Numbers, proposed several new top-level domains like .biz, .aero and .name, which would be managed by other companies and reduce the concentration in one major database like .com. Other companies have already started running their own servers with extensions like .xxx. Not all computers recognize these addresses because most are programmed to start with only the standard top-level domains.

From PC to Shining PC

www.apple.com

1. THE STARTING POINT

A computer linked to the Internet via a dial-up connection or a router on a local area network knows how to send out packets destined for machines elsewhere. It does not know the physical location of the computer it is trying to reach or what path the data should take, but it does know the destination computer's unique Internet Protocol, or I.P., address, and it knows where to hand off these packets to send them on their way.

Routers

I.S.P. NETWORK

2. INTERNET SERVICE PROVIDER

An Internet service provider, or I.S.P., network may have several internal routers that help balance the load among the different companies and individuals it serves. Local and regional I.S.P.'s generally have many connections to higher-level networks run by major network providers, like MCI Worldcom, Cable and Wireless P.L.C. and AT&T.

A

High-level routers

N

3. HIGH-LEVEL ROUTERS

High-level routers at I.S.P.'s and major networks constantly update multimegabyte lists of end-to-end paths linking all the networks on the Internet. These routers also maintain separate lists of the most efficient routes around their own local networks, like MCI Worldcom's national network. The best route may not be the shortest or fastest; it is the route involving the fewest networks.

4. CHOOSING A PATH

High-level routers choose a set of networks for packets to travel across from start to finish. A router on each network receives the packet, then uses the local network's list of routers to hand it off within that network and send it on to the next network on the route. Based on the list it maintains, Router A determines that the packet should cross MCI Worldcom's network and move through a network exchange point to Cable and Wireless, then to Apple's network. MCI Worldcom's routers determine, for instance, that the best path from A to G is through B and D. The whole path from end to end is described as A-B-D-G-L-P-Q.

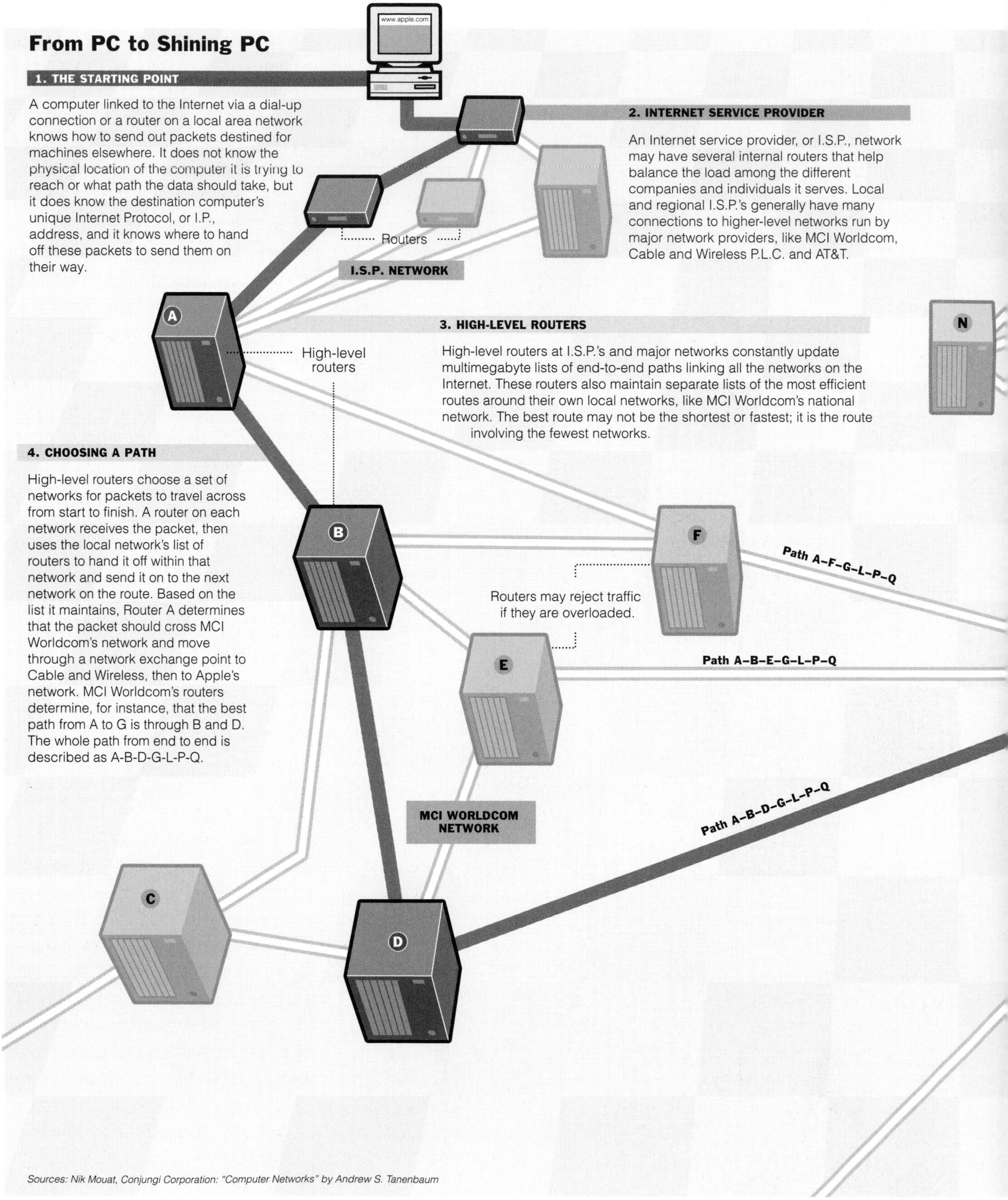

B

F

Routers may reject traffic if they are overloaded.

E

Path A–F–G–L–P–Q

Path A–B–E–G–L–P–Q

MCI WORLDCOM NETWORK

C

D

Path A–B–D–G–L–P–Q

Sources: Nik Mouat, Conjungi Corporation: "Computer Networks" by Andrew S. Tanenbaum

To Sail Data Across the Web, Computers Seek the Best Routes

By Glenn Fleishman

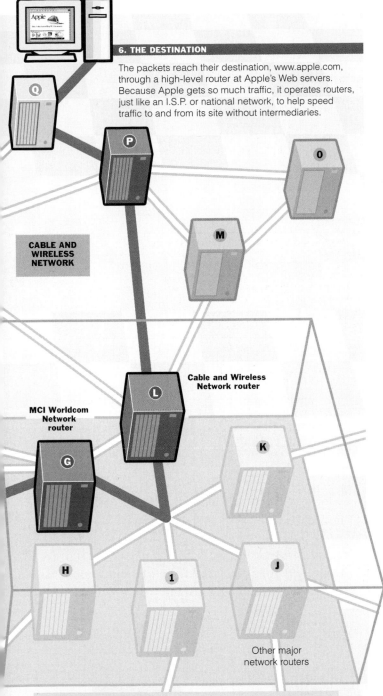

6. THE DESTINATION

The packets reach their destination, www.apple.com, through a high-level router at Apple's Web servers. Because Apple gets so much traffic, it operates routers, just like an I.S.P. or national network, to help speed traffic to and from its site without intermediaries.

CABLE AND WIRELESS NETWORK

Cable and Wireless Network router

MCI Worldcom Network router

Other major network routers

5. NETWORK INTERCHANGE

Regional and nationwide networks rent space and put in routing equipment at several locations around the country, called network access points. Different networks have different policies; they generally allow all networks of the same size to exchange packets of data at no charge at these points but sometimes restrict smaller regional or local I.S.P.'s from having direct access. Many I.S.P.s create private network access points by buying connections from several nationwide networks, allowing them to bypass these sorts of bottlenecks. Larger networks also band together in private exchanges.

Baden Copeland

Billions of times a minute, computers reach out into the maze of machines and connections that form the Internet to find other computers. A bit of magic? Not quite.

Computers connected to the Internet need not know exactly how to reach every other computer. They require only a few simple pieces of information to have their requests or messages wind up at the right machine. Each computer needs to know what its unique number is and how to reach the rest of the Internet. Everything else derives from that.

An Internet Protocol, or I.P., address gives the computer its unique location. The I.P. address, four sets of digits separated by periods (like 192.168.0.1), uniquely identifies both a particular machine and its local network.

A local network is simply a place where all the machines start with the same few numbers. For instance, an address of 192.168.0.1 means that the network is 192.168.0.0 and that the machine's number is 1.

Internet service providers maintain pools of I.P. numbers that they typically assign at random when a user dials up via modem. Every computer connected to the Internet has to have an I.P. number.

Any request for a Web page or mail server that has an I.P. address that starts with numbers that differ from those identifying the computer's local network has to be sent somewhere else. Each Internet-connected computer knows to send all of such requests to the local network's router — a specialized computer that handles the task of sending data among networks.

A router moves data around the Internet along high-speed lines owned by telecommunications companies and network providers. These links connect I.S.P.'s and larger networks, even ones that cross national borders. Small- and medium-size businesses often use a simple router that knows how to send data only up a chain of routers to an I.S.P. or regional or national network. Corporations, major networks, and large I.S.P.'s use high-level routers that keep hundreds of thousands of pathways in their memories at all times.

Computers split data into smaller chunks to share the conduits that make up networks more efficiently. Instead of sending a long, continuous Web page, for instance, the computer breaks up the information into smaller, numbered pieces called packets. Each packet is labeled with the addresses of the sending and receiving computers.

Fresh From Your Server

In the constant back and forth between Web surfers and sites, the bits of data known as cookies can play a big role, providing the site with information about the user and helping track visits. Here's what happens in a typical scenario involving a first visit to a site.

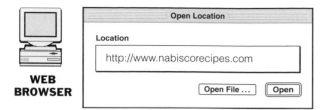

WEB BROWSER

```
Open Location

Location

http://www.nabiscorecipes.com

                    Open File ...    Open
```

❶ **Browser connects to Web site and sends this message:**

GET/HTTP/1.1
Host: www.nabiscorecipes.com:80
Referer: http://www.nabisco.com
Browser: Mozilla/4.0 (compatible; MSIE 4.01; Windows 98)

When a user connects to a Web site that uses cookies — short pieces of information that are stored on users' computers — a set of handoffs occurs between the user's browser and the Web server. The user types in a request for a Web page. Behind the scenes, the browser translates that into a format that the Web server can understand. The browser opens a connection to the Web server and sends as plain text the page request, as well as some information about itself (its platform, version number and operating system).

WEB SERVER

❷ **Server sends response back to browser:**

HTTP/1.1 200 OK
Date: Sun, 14 Jul 2002 JSWF BDAY
Server: Apache/1.3.4 (Unix)
Set-Cookie: visit=1102918597974995;
path=/;
expires=Mon, 15 Jul 2002 09:11:57 GMT
Last-Modified: Mon, 04 Mar 2002 23:57:13 GMT
Accept-Ranges: bytes
Content-Length: 7790
Content-Type: text/html
Web page follows here in HTML
(Hypertext Markup Language)

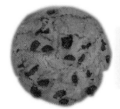

The Web server software usually runs on a computer more powerful than even a high-end home personal computer. A single machine might run software that handles requests for many Web sites. The server software packages its response with header information that helps the browser interpret the page. That includes language and page length in bytes. The server might also send a cookie.

Cookies Leave a Trail of Crumbs Throughout the Web

By Glenn Fleishman

Tell people that you're going to track their every move on a Web site, store that information in files and analyze it later, associating it with personal data they gave the site earlier, and the response might be, "Back off, Big Brother!"

But that is not a paranoid vision of personal-data piracy. It's simply what happens when, as you browse the Web, you (or your browser, without your knowledge) accept a "cookie," a short bit of text that a Web site can store on a user's machine. In other words, it happens every day, millions and millions of times.

The term "cookie" has been used by computer scientists for a long time, but its origin is murky. A Web site uses a cookie to recognize return visitors. It can be no more than 4,096 characters long, but it is often as short as 10 or 20 characters.

Cookies can let users avoid tediously typing in their user names and passwords at sites that require them. And cookies help shopping sites keep track of a limited amount of information, like the contents of a shopping basket or a mailing address.

They are ubiquitous because they smooth the unending stream of transactions between Web browsers and Web servers that make up the constant electronic chatter of the Internet.

When a user types in a Web address or clicks on a link, the browser software sends a request for the Web page to a Web server, another computer with specialized software that receives and processes requests for Web pages, graphics, sounds and other elements. If no cookie is involved, each request for a document or graphic is handled the same way each time. If a cookie is involved, the site knows you've been there before and may know your preferences.

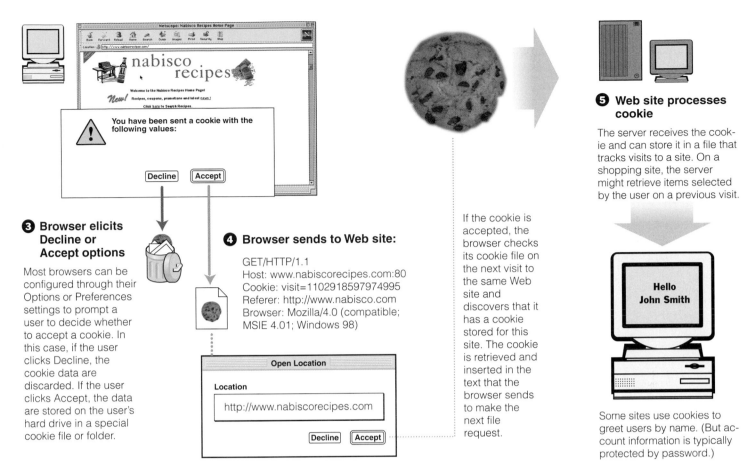

❸ Browser elicits Decline or Accept options

Most browsers can be configured through their Options or Preferences settings to prompt a user to decide whether to accept a cookie. In this case, if the user clicks Decline, the cookie data are discarded. If the user clicks Accept, the data are stored on the user's hard drive in a special cookie file or folder.

❹ Browser sends to Web site:

GET/HTTP/1.1
Host: www.nabiscorecipes.com:80
Cookie: visit=1102918597974995
Referer: http://www.nabisco.com
Browser: Mozilla/4.0 (compatible; MSIE 4.01; Windows 98)

If the cookie is accepted, the browser checks its cookie file on the next visit to the same Web site and discovers that it has a cookie stored for this site. The cookie is retrieved and inserted in the text that the browser sends to make the next file request.

❺ Web site processes cookie

The server receives the cookie and can store it in a file that tracks visits to a site. On a shopping site, the server might retrieve items selected by the user on a previous visit.

Some sites use cookies to greet users by name. (But account information is typically protected by password.)

Baden Copeland

On a site that issues cookies, the software that handles requests from browsers for pages and images can issue a unique identifying number the first time a browser makes a request for pages. That identifying number is sent to the browser, which stores it as a cookie on the user's hard drive.

The next time the user wants to go to the same site the browser sends the identifying number as part of the request. That helps the company that runs the Web server track the number of different visitors to its site. It also means that a record can be kept of all visits by that browser to the site. Cookies can be set to expire at a specific date and time.

Virtually all major electronic commerce sites use cookies to send a browser a session identifier that allows a user to drop items in a shopping basket, browse other sites and return within hours or even days without losing any of those selections.

It may sound spooky, but it's important to recognize cookies' limits. They are not active spying programs, just plain text. A cookie can't suck information from a user's machine and secretly transmit it to a Web server.

A browser delivers a cookie only to the particular site that sent the cookie to the browser in the first place. That keeps one site from raiding another site's cookies. The only information in a cookie is what the user provides. If you don't give information about yourself to a site by filling out a form or otherwise handing over personal details, that site has little or no way to connect you, as an individual, with your surfing expedition through its pages.

One way to deal with cookies is to just say no when a server offers your browser one. All the current versions of Web browsing software offer options in their security preferences, or in a specific cookie-setting panel, for automatically refusing all cookies or for accepting them on a case-by-case basis.

If you refuse to accept cookies, however, some sites won't let you in, others dump your shopping cart each time you leave the site, and still others behave erratically or forget your preferences.

Making Do With Less

The digital information age would hardly be possible without compression — data files would be huge and unwieldy. Although compression can be done in many ways, the goal is always the same: boiling down data to essential information.

Information Compression

One principle of shrinking data to a manageable size is the idea that chunks of information can be represented by symbols that act as a kind of shorthand. Then the shorthand can be decoded to restore the information in the file. One method, the Lempel-Ziv solution, was used for sending telegrams. Common words or phrases could be represented by letters, symbols or numbers, like this:

UNCOMPRESSED:

How much wood could a woodchuck chuck if a woodchuck could chuck wood.

COMPRESSED:

1=WOOD
2=COULD
3=CHUCK

How much 1 2 a 13 3 if a 13 2 3 1.

Image Compression

The most laborious way to store or transmit a digital image is to describe every tiny piece of information, called a pixel, but such files are too large to be handled easily. So an image can be compressed, using math-based methods, before it is stored or sent electronically, then uncompressed when it is opened again. Some data compression methods lose part of the information from an image, while other methods leave an image, like a photograph, unchanged. Here are some examples of compression:

COMPRESSION WITHOUT LOSS OF INFORMATION

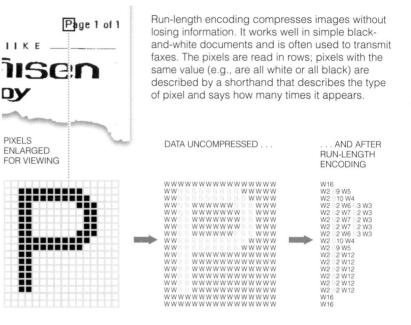

Run-length encoding compresses images without losing information. It works well in simple black-and-white documents and is often used to transmit faxes. The pixels are read in rows; pixels with the same value (e.g., are all white or all black) are described by a shorthand that describes the type of pixel and says how many times it appears.

PIXELS ENLARGED FOR VIEWING

DATA UNCOMPRESSED . . .

. . . AND AFTER RUN-LENGTH ENCODING

Compression: Putting the Squeeze on Data

By Peter Wayner

Most software packages offer their users the chance to use compression to save disk space and make downloads faster. But what is compression? It's a shorthand system that computers use to save space. In other words, it's like lingo, buzzwords or telegram-speak — short abbreviations people adopt to save words.

Understanding "compression" begins with understanding "information" because the two are linked. Computers try to compress files by boiling them down to their essential information.

Many theoretical insights into compression came in the late 1940's when Claude Shannon worked for Bell Labs and developed a theory of information. He created a mathematical model that measured information by its surprise value. Unexpected data has more information in it than data that falls into a regular pattern, he said. One side effect was that this work provided the insight into building optimal mathematical tools, or algorithms, to achieve compression.

While finding that the ultimate compression system is probably impossible, many programmers have come up with fairly good approximations. Most systems scan a file to identify common patterns of letters, numbers or data, then come up with a shorthand system to represent these common patterns in a smaller amount of space. Some of these systems are tuned to particular types of data, like images or video or sound files, and others will work with all forms of data.

Two of the general compression systems that emerged from this era are known as Huffman coding and Lempel-Ziv compression. Many more recent solutions like Pkzip are variations on these algorithms. Instead of describing these more modern systems in detail, it is easier to describe some of their earlier precursors, from the telegraph era, that also used the same basic ideas.

One of the principal notions is to use short code words for the most common patterns and longer ones for the less frequent patterns. That idea was used in Morse Code, which

COMPRESSION IN WHICH SOME DETAILS ARE LOST

A compression method like **JPEG** not only rearranges the image data but also handles the pieces in bigger **BLOCKS** instead of in **SINGLE PIXELS**. It reduces the data density, but that change is not likely to be noticed by the human eye. But if an image is made too large after JPEG compression, the image looks rough.

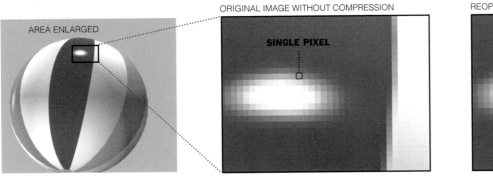

AREA ENLARGED

ORIGINAL IMAGE WITHOUT COMPRESSION

SINGLE PIXEL

REOPENED IMAGE AFTER COMPRESSION

BLOCK OF PIXELS

ANOTHER APPLICATION FOR COMPRESSION: A MOVING IMAGE

A moving image is a sequence of successive frames and can be compressed in much the same way as a still image. But there are also two additional methods: one detects the motion and builds a new frame just by telling where to SHIFT the existing pixels from the previous frame (frames 1 to 2). Another way is to describe only the NEW PIXELS from one frame to another (frames 2 to 3). Both methods can also be used at the same time.

1

2 SHIFT

3 NEW PIXELS

Mika Gröndahl

assigned short codes for common letters (like E equals dot and T equals a dash) and long codes (like P equals dot-dash-dash-dot) for less common letters. In Huffman coding, an option used by most of the major compression software packages, a table is made of the common letter patterns, which are assigned to short patterns of ones and zeros.

Of course, these solutions produce results that depend upon the data being compressed. English text will usually be reduced about 50 percent to 70 percent by Huffman coding, but the results depend upon the writer and the data. If there is plenty of repetition and cliches, then Huffman coding can usually identify more common passages and save more space by assigning shorthand representations for these common phrases. On an intuitive level, it should be easy to see that the less information a document has, the more compression that can be done. Dr. Shannon formalized that idea with mathematics.

Another common technique is known as the dictionary method; the Lempel-Ziv solution is a variation of that general class. Again, this idea was often used when sending telegrams. In the early part of this century, companies competed to create code books for each industry, assigning numbers to common phrases from each type of business.

In the telegram era, both sides needed copies of the same code book. The Lempel-Ziv solution includes a way to incorporate a copy of a code book suited for a particular file into the final compressed version of the file.

Other compression algorithms are tuned to particular types of data. The Joint Photographic Experts Group, or JPEG, solution can compress photographs to one-hundredth of their original size, although the images often lose resolution in the process because some information about the image is lost. The solution trades off compression for image quality. Less compression means that more information is left in the file and less damage is done to the image.

Solutions like JPEG and its moving picture cousin called MPEG are complex and difficult to describe, but a simpler version can explain how some solutions are adapted for particular classes of data. Black and white line drawings, for instance, can often be compressed significantly with a solution known as run length encoding. That consists of going through the image pixel by pixel and counting the number of times that each type of pixel is repeated. A pattern like black black black black white white would be compressed to B4,W2. This works quite well on line drawings with big blocks of black or white because many pixels can be boiled down to one entry.

Squeezing Sound

Compression of digital audio exploits the fact that human hearing is not perfect. For example, loud sounds can mask other sounds, making them inaudible to the human ear. This masking allows the amount of data to be reduced from the original signal while losing little if any sound quality.

Digital audio, like that on CDs, uses only a fraction of the original file size when converted to the MP3 format so the files can easily be distributed over the Internet.

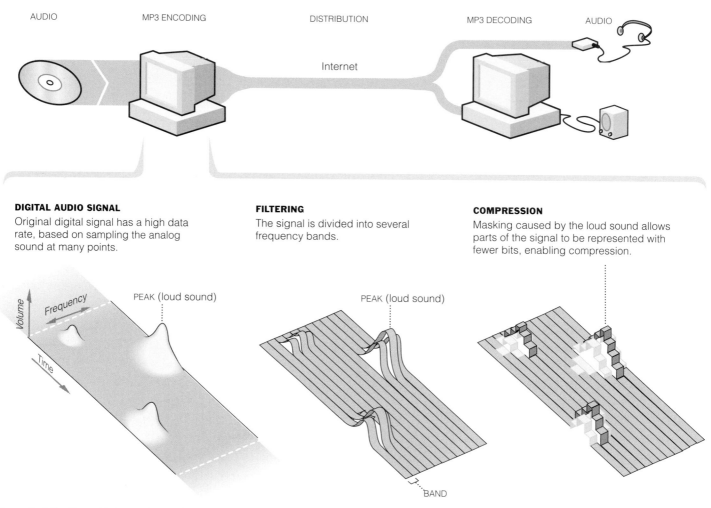

AUDIO   MP3 ENCODING   DISTRIBUTION   MP3 DECODING   AUDIO

Internet

DIGITAL AUDIO SIGNAL
Original digital signal has a high data rate, based on sampling the analog sound at many points.

FILTERING
The signal is divided into several frequency bands.

COMPRESSION
Masking caused by the loud sound allows parts of the signal to be represented with fewer bits, enabling compression.

PEAK (loud sound)

Volume

Frequency

Time

PEAK (loud sound)

BAND

Source: Cambridge Research Laboratory, Compaq Computer Corporation

Mika Gröndahl

An Alphabet Soup of Formats

There are many different file formats for handling audio on computers and the Internet. Here's a sampling.

WAV The standard Windows sound file format, created by Microsoft and IBM. With an encoding rate of about 1,400 kilobits per second, wav files are very large. A three-minute song would take up about 30 megabytes of storage space.

AIFF Short for Audio Interchange File Format, AIFF is to Macintosh what wav is to Windows: a widely used format that features extremely large files.

MIDI Short for Musical Instrument Digital Interface. MIDI files are very small, because they only contain the instructions for creating music. They must be created using a MIDI keyboard or other instrument and software.

MOD Originally developed for Amiga computers. A MOD file includes instrument (or voice, or any sound) samples and instructions on how to play them.

MP3 Short for MPEG-1, Layer 3. Research began in 1987; became part of the MPEG (Motion Picture Experts Group) specifications in 1992. The first portable MP3 player was marketed in 1998. The MP3 format can greatly compress wav or other large audio files without appreciable loss of quality. A three-minute song in MP3 format encoded at 128 kilobits per second takes up less than 3 megabytes of space.

REALAUDIO Often used for streaming audio on the Web, and with software players like RealPlayer. An earlier version, GT, offered compression rates and quality comparable to that of MP3, but with security features so that a song's owner, for example, could restrict downloading. In 2000, RealAudio 8 was released, offering CD-quality encoding at 64 kilobits per second. A three-minute song encoded at that rate takes up less than 2 megabytes.

QUICKTIME Another format that can be used for streaming audio and with a software player. Developed by Apple, but now works with Windows as well.

WMA Windows Media Audio format, released by Microsoft in 1999. Offers quality comparable to a 1 MP3 but at only 64 kbps. WMA files also can be copy protected. Playable with Windows Media Player and many MP3 players that have been upgraded.

Minding Your MP3s: Audio Compression Lets the Beat Go Online

By David Kushner

With all the talk about MP3, the compression software that enables CD-quality music to be sent over the Internet, more and more people believe that MP3 is synonymous with online music.

But just as Kleenex is only one brand of tissue, MP3 is only one major step in the evolution of music online, which began in the late 1980's. The simplest way to describe it is as a way of storing music in digital code. But it is by no means the first way or the last.

Among the first, at least in terms of music online, were WAV and AU files. With the rise in the late 1980's of bulletin board services — electronic message systems formed around common interests — a subculture emerged of fans who swapped songs in one of those formats.

But because of these formats' limited compression technology and the fact that early modems had bandwidths the size of a cocktail straw, a three-minute song could take hours to download at the common modem rate of 2,400 baud.

In 1987, the Motion Picture Experts Group, or MPEG, of the International Organization for Standardization in Geneva began researching and developing a way to compress digital video — in other words, to reduce the amount of code, and thus the amount of storage space, needed for a given image. The result, in 1992, was the MPEG format, which could also be applied to audio files. It reduced storage space without sacrificing CD-quality sound.

MPEG-1 Layer 3, which would become known as MP3, was the most powerful version, capable of squeezing a song into one-twelfth the space that a WAV file would require. With the simultaneous increase in bandwidth to 14.4 and then 28.8 kilobits per second and even faster speeds, the new compression technology meant that the same three-minute song that took hours to send as a WAV file could now be sent in minutes.

Since MPEG was originally intended for digital video, it did not immediately catch on for music. Instead, Web surfers flocked to software like Real Networks' RealPlayer and Microsoft's Net Show, which allow them to "stream" audio quickly (to download songs or other sound in such a way that it can be played as it is being downloaded). Sitting at the keyboard, a user could listen to live concert Webcasts, radio and, yes, pirated recordings without lengthy downloading. But the compromises in sound quality helped create a demand for something that would provide CD-quality sound.

It was not until the mid-90's that college students on higher-speed networks began rallying around MP3, which is considerably more efficient at compressing the space needed for the digital code. With MP3, a dozen albums can be squeezed onto a single CD-ROM.

Like many popular music trends, the grass-roots movement around MP3 soon caught the attention of entrepreneurs — epitomized in the late 1990's by Napster, the service that allowed Web users to download music files from the hard drives of other users. The popularity of the compression format was also fueled by the development of portable MP3 players — smaller-than-a-Walkman devices that can store and play the files — around the same time.

By then, MP3 had caught the attention of record industry executives, who were rocked by the potential invasion of their market and concerned about copyright violations and loss of income.

The Recording Industry Association of America (and the Fraunhofer Institute, the Geneva organization behind MP3) worked to develop what it called the Secure Digital Music Initiative, or S.D.M.I., standard, which would include a way to handle artists' royalties. S.D.M.I. involves a kind of digital "watermarking," inaudible data incorporated into songs that would either allow or prevent a file from being played, depending upon whether a fee had been paid. The idea was for S.D.M.I. to work with existing formats, like MP3, and that new formats could be created using it.

But an early verison of S.D.M.I. never really caught on, and a more complete standard was delayed by bickering between record companies, device manufacturers and software companies.

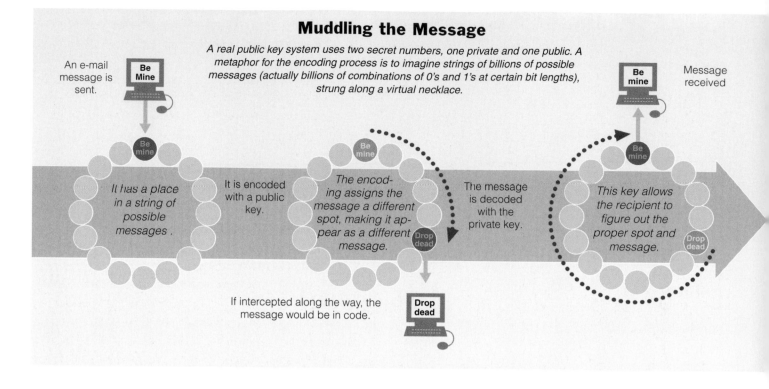

Muddling the Message

A real public key system uses two secret numbers, one private and one public. A metaphor for the encoding process is to imagine strings of billions of possible messages (actually billions of combinations of 0's and 1's at certain bit lengths), strung along a virtual necklace.

An e-mail message is sent.

Be Mine

It has a place in a string of possible messages.

It is encoded with a public key.

The encoding assigns the message a different spot, making it appear as a different message.

If intercepted along the way, the message would be in code.

Drop dead

The message is decoded with the private key.

This key allows the recipient to figure out the proper spot and message.

Be mine

Message received

From Toy Rings to Sophisticated Codes, a Quest for Secrecy

By Peter Wayner

If philosophers taught a class on encryption, they would probably begin by saying: ''Before we ask what encryption is, we must wonder, What is secrecy? What is disclosure? What is information? What is deception? Ultimately we must truly ponder whether the true nature of human existence is to communicate or miscommunicate.''

While it may be unfair to parody philosophers in this way, there is little doubt that the definition of encryption is caught up in the murky depths of knowledge and language. Encryption, or cryptology, is the science of designing code techniques so people can make information unreadable to all but the intended recipients.

Many people's first introduction to the topic were the toy decoder rings that used to be found in cereal boxes. The code rings imitate a system that dates from the time of Caesar. It simply replaces each letter with another letter three or four places down the alphabet. "A" becomes "D," "B" becomes "E" and so on. That is easy to figure out, but everything becomes complicated after that.

The philosophical problems with encryption are debated today at the highest levels of Government, largely because the prevalence of computers lets almost anyone use sophisticated encoding systems. But for most people, the practical question is simply what software can be used to scramble messages reliably so they cannot be read by thieves, stalkers, nosy neighbors, little brothers and the person in the next cubicle.

One way of thinking about encryption is that it involves math problems that can be done forward but not backward. In other words, encryption has to operate on a kind of mathematical frontier. If the methods and problems were well understood, the codes would be easy to break.

At the center of every encryption system is a secret number or mathematical operation, metaphorically referred to as a key. With the toy decoder ring, the key might be the number 3, representing how far down the alphabet to shift the letters. In modern systems, keys are often very large numbers. The mathematical formula for encoding a message — which determines how the key is used — is called an algorithm. For the decoder ring, the algorithm tells you to move a certain number of letters.

One of the most popular algorithms is known as D.E.S., for Data Encryption Standard. It was developed in the 1970's by International Business Machines with some assistance from the National Security Agency. While some people worried that the Government had introduced hidden weaknesses that could

Secrets Kept and Found

The development of secret codes closely followed the development of a standard written language in many civilizations. Here are some high points:

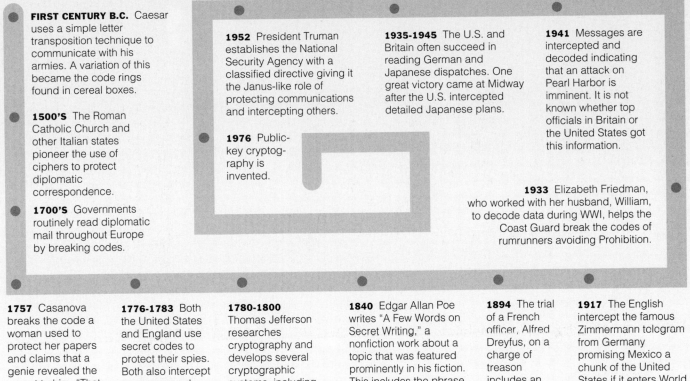

FIRST CENTURY B.C. Caesar uses a simple letter transposition technique to communicate with his armies. A variation of this became the code rings found in cereal boxes.

1500'S The Roman Catholic Church and other Italian states pioneer the use of ciphers to protect diplomatic correspondence.

1700'S Governments routinely read diplomatic mail throughout Europe by breaking codes.

1952 President Truman establishes the National Security Agency with a classified directive giving it the Janus-like role of protecting communications and intercepting others.

1976 Public-key cryptography is invented.

1935-1945 The U.S. and Britain often succeed in reading German and Japanese dispatches. One great victory came at Midway after the U.S. intercepted detailed Japanese plans.

1941 Messages are intercepted and decoded indicating that an attack on Pearl Harbor is imminent. It is not known whether top officials in Britain or the United States got this information.

1933 Elizabeth Friedman, who worked with her husband, William, to decode data during WWI, helps the Coast Guard break the codes of rumrunners avoiding Prohibition.

1757 Casanova breaks the code a woman used to protect her papers and claims that a genie revealed the secret to him. "That day," he wrote in his memoirs, "I became the master of her soul and I abused my power."

1776-1783 Both the United States and England use secret codes to protect their spies. Both also intercept messages and decode them. Several members of Congress, in fact, are skilled in the art.

1780-1800 Thomas Jefferson researches cryptography and develops several cryptographic systems, including the one used to communicate with the Lewis and Clark expedition, 1803-1806.

1840 Edgar Allan Poe writes "A Few Words on Secret Writing," a nonfiction work about a topic that was featured prominently in his fiction. This includes the phrase, "It may be roundly asserted that human ingenuity cannot concoct a cipher which human ingenuity cannot resolve."

1894 The trial of a French officer, Alfred Dreyfus, on a charge of treason includes an encrypted telegram.

1917 The English intercept the famous Zimmermann telegram from Germany promising Mexico a chunk of the United States if it enters World War I. England decodes it and leaks it to the Americans, precipitating the end of neutrality.

Kris Goodfellow

---

later be used to decode messages, none have been found.

D.E.S. is a good method, but it is getting old. Its keys are 7 bytes (56 bits) long. (A bit, in computer memory, is a 1 or a 0.) With enough computer power, it is possible to decode a particular message by testing all possible combinations of 56 1's and 0's — somewhere around 50 quadrillion combinations — to find that message's key.

One way to protect a message is to encrypt it with D.E.S. three times, using three different keys. That would seem to make an attack based upon the brute force of high-powered computing practically impossible, although even this assumption might be suspect. Other ways to achieve complete security include similar algorithms that use substantially longer keys to prevent attack. Some popular names include RC-4, IDEA and Blowfish.

Each of these algorithms is sometimes called a secret-key algorithm or a symmetric algorithm because both the sender and the receiver must agree upon the same key and keep it secret.

Another important technology, developed in the 1970's, is known as public key cryptography. It gets its name because each person has two keys and makes one of them public. Once a message has been encrypted with a public key, only the corresponding private key can decode it.

Public key encryption is popular because it makes it relatively easy to distribute keys. Anyone using a public key can just post it on the Web; several public key servers are already available.

There are several companies that make encryption software for use on a home computer. Many other companies offer products that interact with your e-mail program and make it easier to communicate in private. Some e-mail programs offer built-in encryption as a feature. Most of the practical algorithms have one fundamental problem: there is no way to prove that they really are secure. A math genius may find a way to break the codes. One may already have done so and kept the knowledge secret in order to exploit it.

Chips and Mirrors

Flatbed scanners use a special kind of chip called a charge-coupled device to convert an analog image or document into a digital one that can be stored and manipulated. The charge-coupled device measures light intensity and converts it to a voltage. Here's how a typical scanner works.

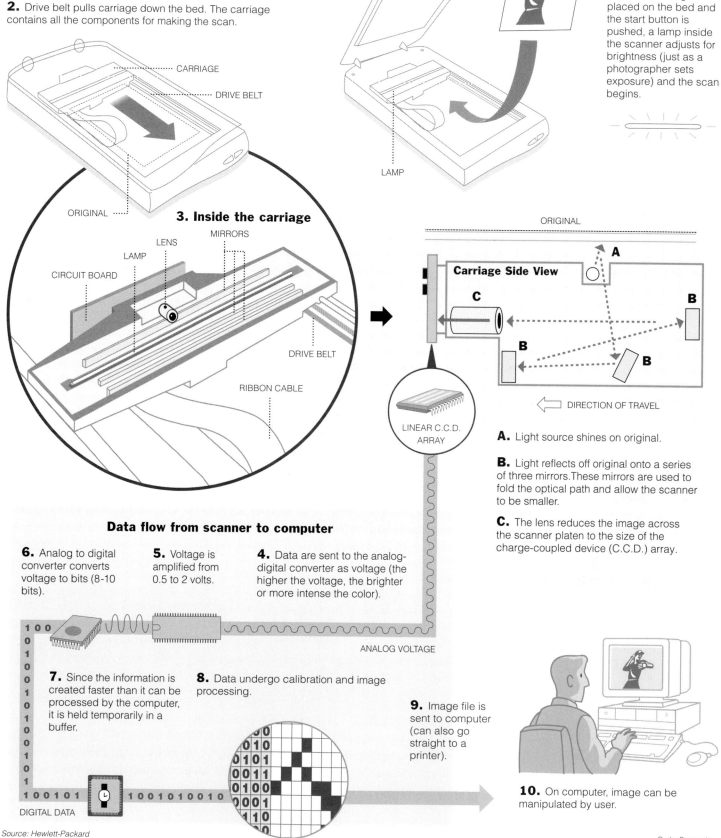

2. Drive belt pulls carriage down the bed. The carriage contains all the components for making the scan.

CARRIAGE

DRIVE BELT

ORIGINAL

1. Once an original is placed on the bed and the start button is pushed, a lamp inside the scanner adjusts for brightness (just as a photographer sets exposure) and the scan begins.

LAMP

ORIGINAL

3. Inside the carriage

CIRCUIT BOARD

LAMP

LENS

MIRRORS

DRIVE BELT

RIBBON CABLE

Carriage Side View

A

B

B

C

B

DIRECTION OF TRAVEL

LINEAR C.C.D. ARRAY

A. Light source shines on original.

B. Light reflects off original onto a series of three mirrors. These mirrors are used to fold the optical path and allow the scanner to be smaller.

C. The lens reduces the image across the scanner platen to the size of the charge-coupled device (C.C.D.) array.

Data flow from scanner to computer

6. Analog to digital converter converts voltage to bits (8-10 bits).

5. Voltage is amplified from 0.5 to 2 volts.

4. Data are sent to the analog-digital converter as voltage (the higher the voltage, the brighter or more intense the color).

ANALOG VOLTAGE

7. Since the information is created faster than it can be processed by the computer, it is held temporarily in a buffer.

8. Data undergo calibration and image processing.

9. Image file is sent to computer (can also go straight to a printer).

DIGITAL DATA

10. On computer, image can be manipulated by user.

Source: Hewlett-Packard

Gorka Sampedro

Flatbed Scanners Convert Ink to Electrons

By David Kushner

Since the paperless office has turned out to be a fantasy, there have to be ways to get all those words and pictures on pieces of paper into a computer. Thus, the scanner.

Technically, a scanner and a digital camera do the same thing. Each takes an image and translates it into digital information that a computer can store. Each uses a number of lenses and an analog-to-digital conversion chip to convert an image into binary bits. The difference is that scanners are made specifically for processing images of documents, not images of people or scenery.

Back in the mid- and late 1980's, when scanners first came to market, they were predominantly the domain of graphic artists and designers. As computer-aided design became part of everyday business, artists needed a way to store and manipulate things like photographs and newspaper and magazine clippings.

The typical scanner mimics the style and shape of a standard photocopying machine. A large "flatbed" surface supports the document that is being reproduced. But the machine produces a digital image instead of a paper copy. Scanners can also be used with optical character recognition software, which reads the image of a page of text after it has been captured by a scanner, then translates the image into the computer code that a word processor can read. Without such software, a page of text is stored in the same way a drawing or photograph is stored, as an image, not text.

The technology used is generally the same from scanner to scanner. A light source shines upon the selected document, and the image is reflected by a series of mirrors. Then it passes through a lens, which reduces the image and focuses it onto an array of charge-coupled devices. A C.C.D., a semiconductor chip that is sensitive to even tiny amounts of light, measures light intensity and converts it into voltage.

This C.C.D. array is part of a circuit board that contains other electronics as well. Through the use of color filters, the information the array stores about the image includes the intensity of each of three colors: red, green and blue. The colors and shapes are translated into different patterns of voltages by the C.C.D. array. Through other chips, the voltage levels, which are analog signals, are given digital values. In the end, that allows the computer attached to the scanner (or the scanner itself, depending on the system) to re-create the image.

Flatbed scanners are the most popular because they provide a clean, efficient way to handle a large volume of documents, like newspaper articles and fragile family photographs.

Sheet-fed scanners are smaller and more compact because the document is fed through a narrow slot in the middle of the device. They are less versatile than flatbeds because a large object may not fit through the slot and a small one might get crumpled. The image quality tends to suffer, but frequent travelers find that it is enough to get the job done.

For people who want more of a hands-on experience, hand-held scanners are the most portable of all. They are most often used to scan small objects, like business cards or newspaper headlines; the photosensitive strips tend to be only a few inches wide.

The most popular home use for a scanner is storing and transferring photographs; it seems that the family album is increasingly ending up online. Of course, there will always be those who find other uses for the devices. Take, for instance, the theater group in the Williamsburg section of Brooklyn that performed a play in which an actress used a handheld scanner to digitize herself onto a screen. The resulting image, needless to say, was grotesque. Perfection in digital imaging, alas, is not always the point.

On the Firing Line

All inkjet printers have a print head with 300 to 600 firing chambers. These can be thought of as tiny nozzles, squirting thousands of droplets of ink per second in a precise pattern to make up the text and images on a page. How those drops are squirted, however, differs between the two basic types of inkjet technologies. The bubble jet, or thermal, type uses heat to create a tiny bubble in the firing chamber that forces out an ink droplet. The other type uses a piezoelectric crystal, which bends when an electric charge is applied, to push the droplets out of the firing chamber and onto the paper.

PRINT HEAD   NOZZLES

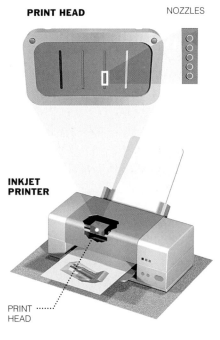

INKJET PRINTER

PRINT HEAD

Software drivers translate an image file from the computer into a pattern of microscopic dots of different colors that blend together to look like a continuous image to the naked eye. That pattern is translated into a sequence of firing instructions for the print head as it moves horizontally in the printer.

Bubble jet printing

RESISTOR

NOZZLE

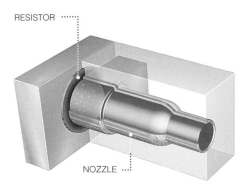

1 To push the ink through the nozzle, bubble jet printers heat minuscule quantities of ink by passing an electrical charge through a resistor, which quickly reaches 900 degrees Fahrenheit.

BUBBLE

INK

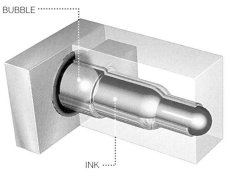

2 The heating element vaporizes a tiny layer of ink at the bottom of the chamber for a few millionths of a second, forming a bubble, pushing the ink down the nozzle.

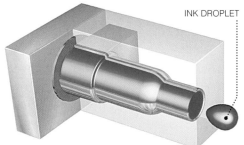

3 The bubble expands and forces a droplet of ink out of the nozzle. Colored droplets are generally so small that a quart could contain 100 billion such drops or more; black droplets are about four times as big. The whole process takes about 10 millionths of a second. The heating element cools and the bubble collapses, creating suction that draws more ink into the chamber from the ink cartridge.

Piezoelectric printing

PIEZOELECTRIC CRYSTAL

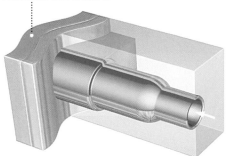

1 A piezoelectric printer works like a squirt gun, but instead of a trigger and plunger, it uses a piezoelectric crystal that changes shape when an electrical charge is applied. A small negative charge deflects the crystal away from the chamber, creating suction.

VIBRATION PLATE

INK

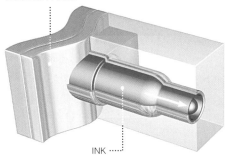

2 A positive charge bends the crystal in the other direction, which pushes a plate into the chamber to create the pressure to expel a droplet of ink.

INK DROPLET

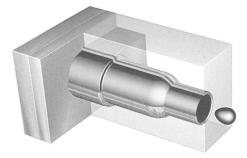

3 An advantage to this approach is that the quantity of ink in the droplet can be precisely controlled. A small charge causes a slight deflection, enough to discharge as little as three-trillionths of a quart of ink through the nozzle. Larger charges can produce larger droplets. The ink is forced out through the nozzle

Sources: Epson America Inc.; Hewlett-Packard

Frank O'Connell

Inkjet Printer Images, Born in a Rainbow of Tiny Drops

By Matt Lake

In the early 1990s, the computer industry gave desktop computer owners a choice of options when it came to printers. But it was pretty much a Henry Ford kind of choice: you could print in any color as long as it was black.

At the time, most people used typewriter-style daisy-wheel printers or clacking dot-matrix printers to put ink on paper. If you wanted a silent printer, your cheapest option was an inkjet machine — and back then, that would have cost you more than $1,000.

Like most computer hardware, inkjet printers have plummeted in price. And as the costs have gone down, the print quality has improved. Many inkjet printers can produce color pages that are almost indistinguishable from photographs. The quality is so high, in fact, that inkjet printers are widely cited as one of the reasons the United States Treasury redesigned its paper currency — it was becoming too easy to print passable counterfeit bills with a good inkjet machine.

It is strange that a technology that produces such sophisticated results began with an accident involving a soldering iron. The story goes that in 1977, an engineer at Canon put a hot soldering iron a little too close to an ink-filled syringe. The heat boiled a tiny amount of the ink in the needle, making it expand into a gas. That pushed some ink out of the tip of the needle.

The physics behind this accident became the foundation of what Canon called bubble jet technology, a printing technique in which electrical resistors heat up tiny bubbles of ink, pushing ink droplets through an array of hair-thin nozzles.

Other inkjet printer makers, including Hewlett-Packard, Lexmark and Xerox, have adapted Canon's thermal technology for their own printer lines, adding enhancements (and patents) along the way.

Another printer manufacturer, Epson America, developed its own variation on the theme. Epson printers use piezoelectric crystals to do the job of the bubbles in inkjet printers. Piezo elements change shape when subjected to an electric charge; they are used in loudspeakers to convert signals from a stereo tuner into sound waves. In Epson's line of inkjet printers, microscopic piezo elements subjected to tiny charges are distorted, pushing out tiny droplets of ink.

Of course, there is more to inkjet printing than the firings of microscopic squirt guns. It is a complicated dance of motors that move a print head with 300 to 600 nozzles across a sheet of paper, squirting 5,000 or more droplets of ink per second in a precise minute pattern.

Most color inkjet printers use four colors of ink: cyan (a pale blue), magenta, yellow and black. They are combined to create all the colors the eye sees in a finished printout. To orchestrate this technique, the computer that is attached to the printer uses driver software that determines the exact color for each dot that will appear on the page. It then figures out how many droplets of each ink color must fall on each dot on the page to make that color. Even when printing at 300 dots per inch (a fairly modest resolution), a letter-size page is more than 2,000 dots wide.

Since photographic prints contain the most complicated shades of color, it is a huge mathematical task for the software to figure out the combination of ink colors needed for each dot. And rendering that subtle color with only four inks can be a slow process.

Despite the precision of halftone software and modern inkjet controllers and print heads, inkjet printers have to contend with one great unknown that can radically affect how good a printout looks: the paper it is printed on.

With the quick-drying inks that Epson, Hewlett-Packard and others use and the tiny ink drops that printers squirt out, it is possible to get decent text and spot colors even on cheap photocopier paper. But big blocks of color, as in photos, can soak and buckle thin paper stock. For tasks like that, the higher contrast and lower absorbency of brilliant white inkjet paper handles color better. And naturally, the best results of all come from glossy photographic paper — which is what you would hope for after paying a premium price for it.

Reading Encoded Data Takes a Light Touch

All compact disks use laser light to "read" the digital information encoded as areas of low and high reflectivity in a pattern on the disk. In a prerecorded CD, a pattern of pits and smooth areas, called lands, is permanently formed in plastic. With CD's that can be recorded once (called CD-R) or many times (CD-RW), the laser that reads the disks also "burns" the patterns.

CD-R
RECORDABLE DISK

CD-RW
REWRITABLE DISK

Data are written on the disk from inside out.

The area nearest the hole, known as the lead-in, contains information on how the disk is read.

LAYER UPON LAYER

A CD is like a sandwich with several layers. The bottom layer is plastic and protects the data; it also contains a spiral groove that guides the path of the laser. In CD-R's, the recording layer is a dye; in CD-RW's, it is a metal alloy, between two dielectric, or insulating, layers. All CD's also have a reflective layer to help bounce the laser light back to a detector.

PROTECTIVE LAYER
LABEL
REFLECTIVE LAYER
RECORDING LAYER (DYE)
CLEAR LAYER
SPIRAL GROOVE TO GUIDE LASER

PROTECTIVE LAYER
LABEL
REFLECTIVE LAYER
UPPER DIELECTRIC (INSULATING) LAYER
LOWER DIELECTRIC LAYER
CLEAR LAYER
RECORDING LAYER

RECORDING DATA ON A CD-R

REFLECTIVE LAYER
RECORDING COLOR
CLEAR LAYER

HIGH REFLECTIVITY   LOW REFLECTIVITY

In a CD-R, when the dye is hit by a laser, that spot heats up and becomes opaque. So when that spot is hit by a laser during playback, the light is scattered. The pattern of opaque and unaffected areas that is created during recording is read by the laser as areas of low and high reflectivity during playback. The dye changes are not reversible.

RECORDING DATA ON A CD-RW

REFLECTIVE LAYER
RECORDING LAYER
CLEAR LAYER

RECORDING LAYER IS ALTERED TEMPORARILY

HIGH REFLECTIVITY   LOW REFLECTIVITY

In a blank CD-RW, the alloy in the recording layer is in a crystalline state. The laser heats the alloy, which then cools quickly, losing its crystalline structure. When the CD is used again, a new pattern of crystalline and noncrystalline (amorphous) areas is created. In playback, the laser reads the crystalline (high reflectivity) and amorphous (low reflectivity) areas.

WHEN IT DOESN'T WORK

Optical disks are vulnerable to scratches and dirt. If a scratch is large enough, it can interfere with the path of the laser light before it reaches the data layer or on its way back toward the detector. Deep scratches can damage the data layer so badly that it becomes impossible for the laser to read the disk.

PLAY SIDE SCRATCH
REFLECTIVE LAYER
DATA
LASER BEAM

LABEL SIDE SCRATCH
LABEL
LASER BEAM

INSIDE THE CD DRIVE

An optical disk is read as it spins by a beam emitted from a small laser. When the beam strikes the disk, it first penetrates clear plastic, which further focuses it. Then it reaches the data layer. Once there, the pattern of reflective and less-reflective areas in the recording layer alters the amount of laser light reflected back to a detector. The pattern is translated into strings of ones and zeros, the digital language that tells the player, in the case of music, to recreate the appropriate sounds.

LASER BEAM
REFLECTION
LENS
LASER
MIRROR
SEMI-REFLECTIVE MIRROR
DETECTOR

Sources: Sony, TDK, *Digital Innovations*

Mika Gröndahl

Recordable CDs: Burning Data With Laser Light

By Michel Marriott

If paleoanthropologists from the distant future ever start digging in your neighborhood, they won't require exotic dating technology to determine what part of the modern age they are rooting around in.

If the scientists discover pie-size, dusty stacks of grooved vinyl platters, they will have stumbled on a lost record store. But if they happen upon scads of shiny five-inch plastic discs in houses, offices, schools, stores, cars and practically everywhere else, they can assume that they have landed in the closing decades of the 20th century, when compact discs reigned.

First developed for the mass market in 1985 by Philips Electronics and Sony Electronics, CDs have rapidly proliferated. After a slow start, they quickly became the medium of choice for music companies and listeners because of their large storage capacity, ease of use and high-fidelity sound reproduction.

By the early 1990's, optical discs had become popular among computer software publishers looking for better ways to store computer programs and other data, leading to the creation of the first CD-ROM (read-only memory). Since then, the number of uses and types of optical discs have exploded. There is CD-R (recordable once), CD-RW (recordable many times), DVD and DVD-ROM (very high-capacity CD-ROMs that are used mostly to view movies), DVD-RAM (recordable DVDs) and minidiscs (three-inch recordable CDs).

The appeal of the optical disc is obvious, said Raymond C. Freeman Jr., who manages the Optical Storage Technology Association and is president of his own data storage marketing and analysis company in Santa Barbara, Calif. "People feel very comfortable using them, even if the people are nontechnical," he said. "You pretty much put the disc in a drive and it works." Well, most of the time.

While optical discs are far less vulnerable to wear and tear than vinyl records, optical discs can become scratched and dirty enough to malfunction. For audio CDs, that usually results in a kind of sonic hiccup, the digital equivalent of a skip on a vinyl record. With multimedia discs, such problems can make it impossible to read the disc at all.

To understand what happens when optical discs do not work, it is helpful to know what happens when they do. A laser head, usually mounted on a track, moves a laser beam in micron-tiny increments a small distance from the face of the disc, which, in the case of music CDs, is spinning at a rate of 200 to 600 revolutions per minute. CD-ROMs spin much faster.

The disc's speed is automatically adjusted to keep the reading speed constant as the laser starts reading at the inside of the disc and reaches the outer edges.

What this finely focused laser beam "sees" on a typical music CD — the kind bought in a music store — is a series of pits and flat surfaces called lands. The microscopic pits and lands, which are pressed into the plastic disc during manufacturing, are arranged in a tightly packed spiral; for a CD at full capacity, or 650 megabytes of data, the spiral, if unraveled, would be a thread 200 times as thin as a human hair and more than five miles long.

When the laser strikes a land, the beam bounces back to its source and is picked up by a detector. When the laser strikes a pit, the light is diffused and returns to the detector at a significantly decreased intensity. As the laser tracks, the detector reads these rapid flashes of high- and low-intensity signals as offs and ons, or zeros and ones. That binary information is translated by a microprocessor back into music, text, graphics or whatever else had been encoded onto the disc.

Home-recordable CDs work on the same basic principle, but the pits and lands aren't permanently pressed in plastic. Rather, dyes or metallic compounds are modified by the laser during recording, creating the equivalent of pits and lands.

CDs are designed to have their data oversampled — read many times — to help correct for errors. But they still skip when they are dirty or damaged. A skip can also be caused by a bump that knocks the laser head off track, much as a phonograph needle skips across a record when it is bumped. Many portable CD players are equipped with what is called shock-resistant memory, a kind of data reservoir that stores what the laser reads. When the laser is disrupted, the player dips into its reserve memory so the listener never hears the break in the data stream.

Or at least most of the time.

A Message's Ups and Downs

In a paging system, a simple message can take a circuitous route involving ground- and space-based transmitters and receivers. In the system shown here — the Skytel network, one of the biggest — each type of message follows a slightly different path, depending on whether it originates from a telephone, a computer or another pager, and the person receiving the message has the option of sending a response to the sender. Each pager has an identification number so the network can find it. The pager periodically sends out radio queries to find the nearest receiver and report its position.

SATELLITE UPLINK

Messages of all sorts are received at the satellite uplink station in Chicago and sent as radio transmissions to a satellite in a geostationary orbit (which means that its orbit matches Earth's rotation, keeping it over the same spot on the ground).

Starting Points

TELEPHONE

The message can be sent several ways — as voice mail, perhaps, or as an alphanumeric message like a phone number.

Hard-wired through a telephone line (unless sent from a cellular phone).

COMPUTER

Via e-mail.

Hard-wired through an Internet connection.

NETWORK OPERATIONS CENTER

The brains of the paging network. Skytel's main center is located in Jackson, Miss. The center is fully automated and contains about 150 computers. From here, all messages are sent to the satellite uplink.

PAGER

Via an alphanumeric message.

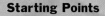

Transmitted through a paired channel in the frequency bands of 901 to 902 megahertz and 940 to 941 megahertz.

RECEIVER

Skytel has several thousand radio frequency receivers positioned around the country, each with a range of approximately 2.7 miles, depending on the terrain. Buildings or mountains tend to limit the range. From here the message is sent on to the network operations center.

Source: Skytel

Mika Gröndahl

SATELLITE
Messages are received by a satellite and transmitted back to Earth. A second satellite is used as a backup.

SATELLITE DOWNLINK, RADIO TRANSMITTER
Downlinks are radio receivers; each one is hard-wired to one of 900 radio transmitters across the country. Each transmitter has an approximate range of 8.8 miles, depending on the surrounding topography.

As pagers automatically send radio signals to the receivers in their area, the network notes which pagers are nearby and transmits messages designated for them.

PAGER
The message is received, and the recipient can then reply.

Replying to a Message

A person who has received a message on a pager can reply to the sender directly by typing in a text message on the pager. The reply travels from the pager to a receiver, then to the network operations center, which figures out the routing.
- If the message is going to another pager, it is sent to the satellite uplink in Chicago, then follows the ground-satellite-ground path to the other pager.
- If it is going to e-mail, it is transmitted through an Internet connection.
- If it is going to a telephone, the text is translated into a digitized voice message, which is sent via the public telephone system.

With Pagers, You've Got The World on Your Belt

By David Kushner

In the digital world, being out of reach is the ultimate faux pas; people have come to expect instant access not only to information but also to one another.

As a result, business cards are overflowing with addresses and numbers for snail mail, telephones, cellular phones, Web sites, e-mail and, of course, pagers. Tens of millions of portable electronic paging devices help to keep people connected worldwide.

If humans are to become cyborgs one day, then the ubiquity of pagers represents some kind of evolutionary step. It's the next best thing to having a tracking chip soldered into your skull. Thanks to an elaborate network of transmitters, receivers and satellites, you can almost always be found if you wear a pager.

This technology has been decades in the making. When pagers first appeared, in the early 1970's, they were designed for receiving messages within a city. To reach someone, a caller had to dial a number and tap in a numeric message. The person receiving the message was notified by a beep and had to call in to get the information — not very efficient.

In the early 1980's, the Mobile Communications Corporation of America introduced alphanumeric paging. The problem was, it didn't represent a direct channel of communication; an operator had to take the call, type in the message and send it on its way.

Today such methods seem as antiquated as Morse code. And although cell phones have all but replaced pagers for some uses, paging culture remains strong. Systems allow callers to send voice messages (which can be converted into text), alphanumeric commands, faxes or e-mail.

Pager companies are continually adding features, like "reflex" technology, which allows two-way communication. Instead of just sending someone a message and hoping that it is received, a sender can immediately get a response from the person with the pager.

All this networking, of course, comes with one unavoidable consequence. The easier it becomes to get in touch, the harder it is to get away.

Reading as It Writes

There have been numerous attempts to link the pen to the computer, to convert handwritten messages to a usable digital form. Here's a look at one effort, the Anoto pen, which contains a tiny camera, a microprocessor and other electronics to read a special pattern of dots as it writes.

1 GATHERING THE INFORMATION

As you write with an Anoto pen, a lens focuses the stationery's pattern onto a light sensor that reads the pattern at a rate of 100 frames per second. It sends the data to the pen's central processor. At the same time, a force-sensing resistor embedded near the pen's ink cartridge measures the force applied to the pen and feeds that information to the processor so the pen analyzes movements only while it is writing.

2 PROCESSING THE INFORMATION

The image processor calculates what the pen is writing by the changes it reads in the pattern. Since it receives 100 images per second, even fast writing can be analyzed. Then the information is passed to the pen's memory chip, which can store several pages' worth of writing.

3 TRANSMITTING THE INFORMATION

The information is sent wirelessly to a compatible cell phone or other receiver. From there it goes to a destination that the writer designates by checking appropriate boxes on the page. The data can show up as a message on a cell phone screen, as a fax or an e-mail note, or in a destination computer via the Internet.

INK CARTRIDGE AND FORCE SENSOR

A sensor near the ink cartridge senses whether the pen is being pushed against the paper. That helps the processor tell if the pen is being used for writing.

A RANGE OF FUNCTIONS

Patterned paper can be printed with boxes that, when checked, tell the Anoto system what to do with the information it collects.

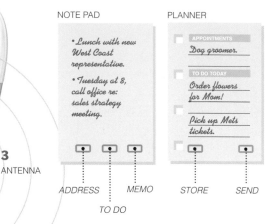

NOTE PAD

- Lunch with new West Coast representative.
- Tuesday at 8, call office re: sales strategy meeting.

ADDRESS | MEMO
TO DO

PLANNER

APPOINTMENTS
Dog groomer.

TO DO TODAY
Order flowers for Mom!

Pick up Mets tickets.

STORE | SEND

3 ANTENNA

TRANSMITTER
A low-power transmitter in the pen has a range of about 30 feet. It transmits in the 2.4-gigahertz band, which is used by industrial, scientific and medical applications.

RECHARGEABLE BATTERY

MEMORY MODULE
The pen has enough memory to store several pages of writing.

2 IMAGE PROCESSOR

CAMERA
The camera uses CMOS technology, which makes it possible to combine sensing and image-processing functions on a single chip. The ink from the pen is not visible to the camera, which reads only the dots to get its positioning information.

PAPER PATTERNS

The Anoto system depends on paper or other surfaces printed with a tiny pattern of dots, which make the surface appear light gray. The pattern looks regular under a magnifying glass, but the dots are slightly displaced from a regular grid. This displacement varies continuously, so that as the pen moves, the system can tell exactly where in the overall pattern the pen is. The system can then re-create the pen's movements.

Not to scale

CHECK BOX

DATE BOOK

Sources: Anoto; Ericsson

Frank O'Connell

A Pen for Scribbling Your Way Across the Internet

By Matt Lake

Not too long ago, if you wanted to send e-mail across the Internet, you needed a computer and a modem. In the relentless move to simplify technology and make it more accessible, a wave of more familiar Internet devices followed, starting with WebTV, which turned your television into a PC screen, and including cellular phones and dedicated e-mail stations that you plug into a phone line.

But if you want a really familiar interface with the Internet, how about putting pen to paper? Among the promising technologies for using a pen to send e-mail, faxes and e-commerce orders was the Anoto Bluetooth pen, developed by three companies: Ericsson, Anoto and Time Manager.

The device looks, handles and writes like an honest-to-goodness ink pen, a bulky one, true, and one with a little L.E.D. indicator light on the side, but an otherwise ordinary-looking writing instrument. But besides the usual ink cartridge, the Anoto pen contains image processing and radio broadcast circuitry designed to transmit what you write, using the Bluetooth wireless networking standard. If the pen is within 30 feet of its owner's Bluetooth-capable cell phone, handheld computer or network base station, it will automatically transmit the information.

There is another element to the Anoto system that enables the whole thing to work properly, and its technology that is hundreds of years old: pen on paper. While the pen can write on any regular paper, it needs special paper to transform itself into a wireless communication system.

A pattern too small to be seen with the naked eye is printed on Anoto stationery. The paper looks ordinary, although with a slight gray hue. But with a magnifying glass, a tiny pattern of dots is visible. It is this pattern that the pen's camera chip picks up and processes at a rate of 100 frames per second. The pen's image processor picks up the changes in the pattern as the pen moves across it. That enables the processor to see the direction and speed in which the pen is moving, and that lets it build an image of whatever letters or symbols are being written down. In that way, the handwritten notes are compiled in the pen's processor.

At that point, the image is sent by the pen's radio transmitter to a compatible cell phone, hand-held device or computer. The pen adheres to the Bluetooth standard, a wireless communication technology developed by a consortium including Siemens, Intel, Toshiba, Motorola and Ericsson. Bluetooth is named for a Danish king from the turn of the last millennium, Harald Bluetooth, whose reign unified Denmark and Norway, and it reflects the Bluetooth Special Interest Group's hope to unify computing and wireless telephony platforms.

With its widespread acceptance among cell phones and a transmission range of up to 30 feet, Bluetooth has an edge over infrared devices that are often used for communicating between two computing devices. Another plus is that it lets devices communicate even when they are separated by objects or walls.

But as with all technologies, success is really in the applications. And with the Anoto pen, the applications are all in the stationery. The patterns of tiny dots form a code that tells the pen what to do with the data it is writing. The Anoto stationery is printed with boxes for faxes or e-mail. If you jot down a note and check one of the boxes, then write the destination phone number or e-mail address, the pen will beam instructions to your cell phone to deliver the note to that number or address. If you make a to-do list for yourself on paper, the pen will beam it to your notebook PC. And if a forward-thinking company sends you Anoto-based order forms, you can shop online with a few flicks of a pen. The Anoto system could also be used to enter Chinese or other keyboard-unfriendly lettering into computers.

One thing is certain: if you have an Anoto in your pocket and someone asks to borrow your pen, make sure that you have a cheap ballpoint to hand over. With all that pricey electronic circuitry inside, the radio pen is one writing instrument you will want to keep chained to your pocket protector.

Entertainment and Sports

A "Black Box" for the Bicycle

Serious bicyclists rely on bicycle computers to determine how fast and far they have ridden. The more sophisticated of these devices, like the SpeedZone P. Brain, can also measure heart rate and elevation, recording the information at intervals for downloading later to a PC for analysis.

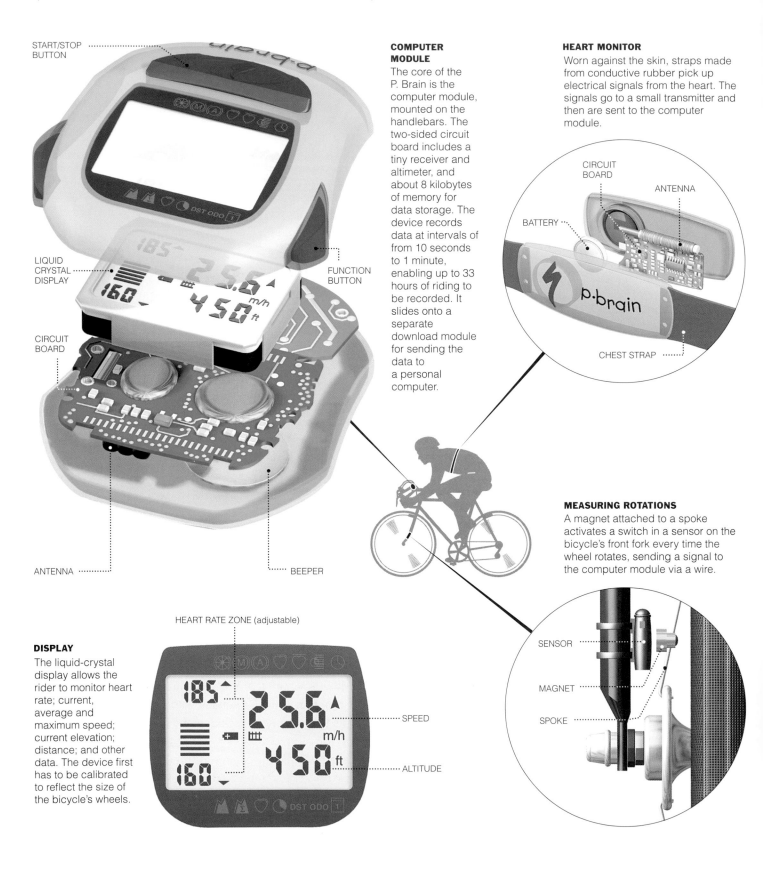

START/STOP BUTTON

LIQUID CRYSTAL DISPLAY

FUNCTION BUTTON

CIRCUIT BOARD

ANTENNA

BEEPER

COMPUTER MODULE
The core of the P. Brain is the computer module, mounted on the handlebars. The two-sided circuit board includes a tiny receiver and altimeter, and about 8 kilobytes of memory for data storage. The device records data at intervals of from 10 seconds to 1 minute, enabling up to 33 hours of riding to be recorded. It slides onto a separate download module for sending the data to a personal computer.

HEART MONITOR
Worn against the skin, straps made from conductive rubber pick up electrical signals from the heart. The signals go to a small transmitter and then are sent to the computer module.

CIRCUIT BOARD

ANTENNA

BATTERY

CHEST STRAP

MEASURING ROTATIONS
A magnet attached to a spoke activates a switch in a sensor on the bicycle's front fork every time the wheel rotates, sending a signal to the computer module via a wire.

SENSOR

MAGNET

SPOKE

DISPLAY
The liquid-crystal display allows the rider to monitor heart rate; current, average and maximum speed; current elevation; distance; and other data. The device first has to be calibrated to reflect the size of the bicycle's wheels.

HEART RATE ZONE (adjustable)

SPEED

ALTITUDE

Source: Specialized Bicycle Components Inc.

Frank O'Connell

The Bike Computer: Coxswain and Trainer in a Small Package

By Glenn Fleishman

When you set out for a leisurely country bike ride, knowing the distance you've traversed to the hundredth of a mile probably isn't your first thought. And you might not want to know how much your heart rate increased during the 10 minutes you spent climbing a hill.

But dedicated cyclists can be obsessive about their performance. And the only way to get an accurate reading on how fast the bike is going or how hard the rider is working is to use a bicycle computer.

Bike computers (sometimes called cyclocomputers by manufacturers and stores) can do more than monitor speed and distance; the most sophisticated can also count the rider's heart rate and determine elevation. These measurements, taken together, offer a series of snapshots of a ride.

The practice of measuring how far and fast a bicycle travels emerged alongside the modern bicycle in the 1860's. Russell Mamone, a past editor of The Wheelmen - a publication of an organization of the same name that is devoted to bicycle history - said that the earliest "cyclometers" had been attached inside front wheel hubs, to the underside of pedals or to wheels, where they counted off rotations by striking pins against the wheel or spokes. None of them could be read without dismounting from the bicycle.

"People in those days were very interested in how far they traveled," Mr. Mamone said. "You couldn't really tell how far it was from Washington to Fredericksburg, Va. To record mileage accurately, they would use these cyclometer records."

Modern bike computers have not strayed too far from the basics. They employ a magnet attached to a front wheel spoke to signal rotations. A matching magnetic sensor fits on one of the blades of the bike's front fork. The magnet and sensor have to be carefully placed so the gap is small enough to trigger the sensor, but big enough to keep the two from knocking together.

Bicycle computers can also calculate cadence — the pedaling rhythm, measured in rotations per minute — by mounting sensors on the back of the bike to track rotations of the bicycle crank, the pedals' pivot point and the place where the gears are mounted. Many biking experts recommend a regular cadence for better performance and exercise value. While it can be hard to gauge r.p.m.'s mentally, a bike computer can provide sound and visual cues.

When a magnetic sensor is used to determine speed or cadence (some can measure both), the magnet induces a current in a coil in the sensor each time it rushes by, and that pulse is sent to the bike computer. That connection can be through a wire run from the sensor to the computer, mounted on the handlebars, or a low-power transmitter on the sensor can send a signal wirelessly.

Before a bike computer can translate the pulses arriving from a sensor into a readout of miles or kilometers per hour, the computer has to be programmed with the wheel's circumference so it can calculate both speed and distance. Bike computers come with settings for standard wheel sizes. But those generally include only a few combinations, so many riders have to make their own calculations. (Measuring the wheel's circumference is easily done by rolling the wheel along a tape measure on the floor.) Some computers can store multiple wheel sizes so adjustments can be made when a rider changes bikes or swaps wheels on the same bike.

With the wheel size in mind, the bike computer can translate the number of rotations into a speed calculation. It counts the number of rotations and multiplies that by the distance traveled with each rotation. Most bike computers can store the distance traveled in the current ride and the total number of miles on the bike, as well as the current, minimum and maximum speeds. Some can also monitor how much time or distance remains in a programmed ride, or how close the rider is to an ideal cadence.

An Electronic Eye Makes the Call

A tennis ball can seem like just a blur to the umpire who must judge whether a booming serve is in or out. At major tournaments like the United States Open, an infrared system called Cyclops by Carlton makes the call on serves that are close to the back line of the service box.

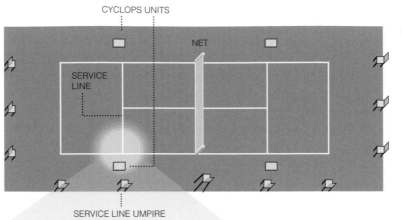

CYCLOPS UNITS

NET

SERVICE LINE

SERVICE LINE UMPIRE

On the court

The Cyclops system consists of a transmitter box and a receiver box, bolted to the court on the sidelines. The boxes are carefully aligned with the back line of the service box. There is an identical setup on the other side of the net.

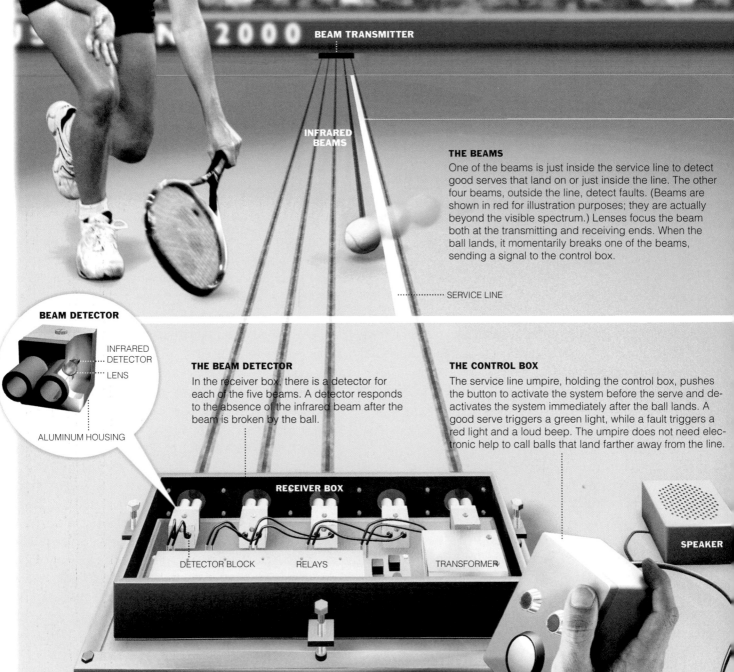

BEAM TRANSMITTER

INFRARED BEAMS

THE BEAMS
One of the beams is just inside the service line to detect good serves that land on or just inside the line. The other four beams, outside the line, detect faults. (Beams are shown in red for illustration purposes; they are actually beyond the visible spectrum.) Lenses focus the beam both at the transmitting and receiving ends. When the ball lands, it momentarily breaks one of the beams, sending a signal to the control box.

SERVICE LINE

BEAM DETECTOR

INFRARED DETECTOR

LENS

ALUMINUM HOUSING

THE BEAM DETECTOR
In the receiver box, there is a detector for each of the five beams. A detector responds to the absence of the infrared beam after the beam is broken by the ball.

THE CONTROL BOX
The service line umpire, holding the control box, pushes the button to activate the system before the serve and deactivates the system immediately after the ball lands. A good serve triggers a green light, while a fault triggers a red light and a loud beep. The umpire does not need electronic help to call balls that land farther away from the line.

RECEIVER BOX

DETECTOR BLOCK   RELAYS   TRANSFORMER

SPEAKER

Source: Sarah Gauci Carlton

Mika Gröndahl

The Tennis Line Monitor, One More Ace on the Court

By Catherine Greenman

Tennis fans who dislike the temper tantrums thrown by some top players over line umpires' calls can thank Bill Carlton that there aren't more of them.

Mr. Carlton, an inventor from Malta, developed a system called the Cyclops by Carlton, a service line monitor that uses infrared beams to determine whether a serve is in or out.

The device, which was developed in the late 1970's and has been refined continually since then, has been used in many major tournaments, including Wimbledon and the Australian and United States Opens. Line umpires say the system helps relieve some of the pressure when calling serves.

"As an umpire, I loved it," said Jay Snyder, tournament director at the United States Tennis Association. "Probably the most argued-over call in tennis is the serve because it's so split-second. But the players don't argue with the Cyclops. They've come to accept it."

Although there are no published accuracy statistics, Mr. Snyder said that Cyclops's calls were accurate more than 99 percent of the time, barring extreme temperature changes that can affect the court surface. Other umpires appreciate the system's objectivity in situations where they could be accused of favoritism.

"The machine doesn't care whether you're from Spain or from Los Angeles," said Brian Early, a referee with the United States Tennis Association. "It just calls the balls the way it is, and that's fine."

Before the United States Open gets under way each year at the National Tennis Center, Bill Carlton's daughter, Sarah Gauci Carlton, arrives from Malta to set up Cyclops systems on center court and on three others where marquee matches are played. Ms. Carlton stays through the tournament to maintain the devices and make adjustments.

The Cyclops system consists of two boxes bolted to the court surface on the sidelines, in line with the back of the service box. One of the Cyclops boxes transmits five infrared beams across the court to the other box, which is the receiver and is connected to a smaller control box held by the service line umpire. There is an identical setup on the other side of the net.

The boxes are positioned so one infrared beam runs along the good side of the service box line, and the other four run on the fault side, up to 18 inches from the line. When a ball hits on the line (or just inside the line) it momentarily breaks the first beam and turns off the others. If the serve is long, it breaks one of the four other beams.

The control box signals with a green light when the serve breaks the first beam. If it breaks any of the others without breaking the first beam, it signals with a red light and a loud beep. A serve well beyond or in front of the beams is considered easy enough for the umpire to judge by eye.

The Cyclops can be used only at the service line because, unlike the baselines or sidelines, there are no feet or rackets present to break the infrared beams. The system must be inactivated (the umpire pushes a button on the control box) immediately after a good serve so the ball, racket or feet do not trip the beams again, triggering more beeps.

Over the years, the Cyclops system has been redesigned to be less sensitive to insects attracted by the lights during night games, Mr. Snyder said. And the whistle that was initially used to indicate a fault has been replaced by a lower-pitched beep, partly because it can be heard better in large stadiums and partly because some spectators would occasionally mimic the whistle sound when they thought that a ball was out.

As precise as it is, the Cyclops system is used to complement, not replace, people. The line umpire has the final call and can shut down the system if there are questions about its accuracy. That rarely happens. At the end of each match of the Open, Ms. Carlton speaks to the service-line umpire about how well the system worked. "There are a few linesmen who are more particular than others, but very few who don't like it," she said.

Just Add Air and Water

Machine-made snow is like the real thing (just a lot more expensive). There are two types of snow-making machines — air guns that use compressed air to break a stream of water into droplets, and fan guns that rely on a high-powered fan for most of the air supply. Air guns require that water and air lines be run along ski slopes, and use large (and costly) compressors and pumps. Fan guns require only water lines, but the guns themselves are much more costly. Here's a look at a fan gun.

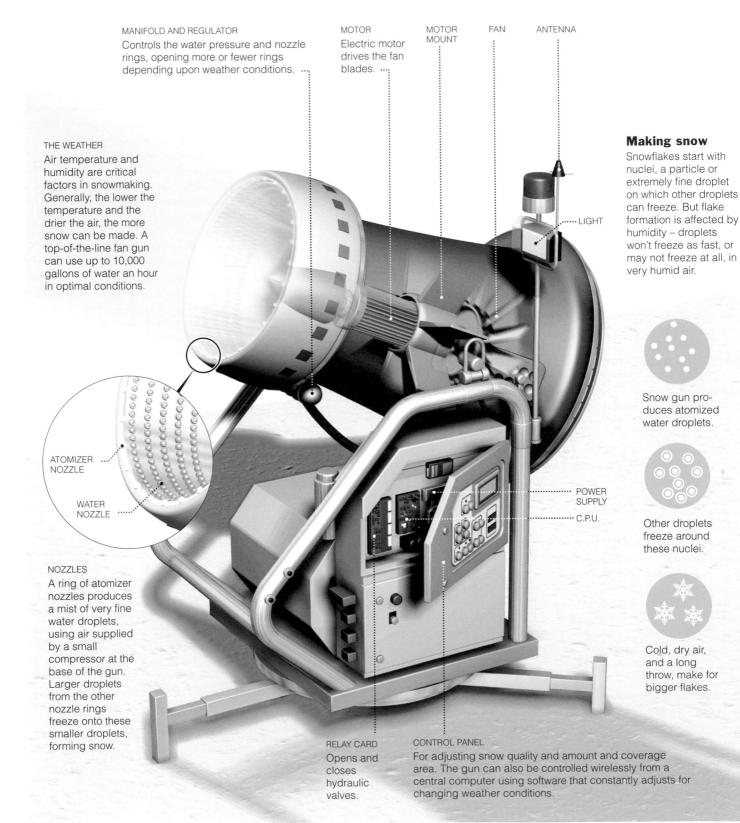

MANIFOLD AND REGULATOR
Controls the water pressure and nozzle rings, opening more or fewer rings depending upon weather conditions.

MOTOR
Electric motor drives the fan blades.

MOTOR MOUNT

FAN

ANTENNA

THE WEATHER
Air temperature and humidity are critical factors in snowmaking. Generally, the lower the temperature and the drier the air, the more snow can be made. A top-of-the-line fan gun can use up to 10,000 gallons of water an hour in optimal conditions.

LIGHT

Making snow

Snowflakes start with nuclei, a particle or extremely fine droplet on which other droplets can freeze. But flake formation is affected by humidity – droplets won't freeze as fast, or may not freeze at all, in very humid air.

Snow gun produces atomized water droplets.

Other droplets freeze around these nuclei.

Cold, dry air, and a long throw, make for bigger flakes.

ATOMIZER NOZZLE

WATER NOZZLE

NOZZLES
A ring of atomizer nozzles produces a mist of very fine water droplets, using air supplied by a small compressor at the base of the gun. Larger droplets from the other nozzle rings freeze onto these smaller droplets, forming snow.

POWER SUPPLY

C.P.U.

RELAY CARD
Opens and closes hydraulic valves.

CONTROL PANEL
For adjusting snow quality and amount and coverage area. The gun can also be controlled wirelessly from a central computer using software that constantly adjusts for changing weather conditions.

Source: Lenko

Frank O'Connell

Snow-making Machines Ease the Economic Bumps for Ski Resorts

By Jeffrey Selingo

In the past, when dry weather lingered in the Northeast in midwinter, that would have meant the cancellation of many a ski trip. But nowadays, even if there is no hint of snow in the forecast, thousands of skiers make their way to mountain resorts, comforted by ski reports on the radio announcing that all the trails are open.

They have snow-making to thank.

Mild winters with little snow once meant financial disaster for the ski industry. But in the past 30 years, the widespread use of snow-making equipment has provided so much stability to a ski resort's season that even traditionalists in the Swiss Alps and British Columbia have decided to install it.

While natural snow is preferable for most resort owners, nature doesn't provide enough of it in many parts of the world for resorts to open early in the season and survive to early spring. "It used to be that the ski industry was satisfied with the economic performance of natural snow," said Curt Bender, a professor of ski area operations at Colorado Mountain College in Leadville, Colo. "But staying open 70 days a season is no longer acceptable. Nowadays, skiing is more of a business, and resorts are required to have that guarantee that they'll perform."

Machine-made snow has become so popular among some resort owners that they sometimes look at nature as getting in the way, especially when it keeps skiers from getting to the mountain. "There's no question that a few inches of snow in Boston is a huge shot in the arm for resorts in Maine, New Hampshire and Vermont," said Skip King, a vice president at American Skiing Company, the largest operator of ski resorts in the country. "But it's less critical than it once was. People know we can make snow."

Still, making snow is not cheap. It is typically the second-largest cost, after labor, for resorts. New systems can cost upward of $40,000 per acre to install and $1,700 per hour to operate.

The earliest snow-making machines used leftovers from household plumbing supplies and irrigation equipment. Although the technology has improved, the basic concept of snow-making has remained the same. Making artificial snow mimics what happens in nature, but the snow is produced much more quickly. In one method, water is pumped to a snow gun, where compressed air blows the stream of water into tiny particles that are then shot out into the air, where they crystallize and fall to the ground as snow.

Walter R. Schoenknecht, owner of Mohawk Mountain in Cornwall, Conn., was one of the first to turn to technology to make an end run around nature in the winter of 1949-50. With no natural snow in early January, Mr. Schoenknecht trucked in blocks of ice, broke it up with an ice crusher and spread the chips over one slope.

Machine-made snow was first mass-produced by three engineers — Art Hunt, Dave Richey and Wayne Pierce — on March 14, 1950, in Milford, Conn. Using a garden hose, a 10-horsepower compressor and a spray-gun nozzle, they produced about 20 inches of snow. They later patented their invention.

"The first attempts in snow-making were oddball solutions to a lack of snow," said Jeffrey R. Leich, executive director of the New England Ski Museum in Franconia, N.H. Despite those early efforts, the widespread installation of snow-making machines did not start until the 1970's, he added.

"For the most part, there was enough snow," Mr. Leich said, "and it wasn't seen as a competitive advantage to have snow-making machines. Now, it's unthinkable that you operate without snow-making."

Making snow is not as simple as running water through a hose. A lot depends on the weather, and it is not the temperature on your outside thermometer — known as the dry bulb temperature — that is important. What snow-making operators look at is the wet-bulb temperature, which is adjusted for humidity.

When the humidity is high, water may not freeze even at temperatures lower than 32 degrees. And when the air is extremely dry, water can freeze even at temperatures higher than 32 degrees. So low air temperatures and low humidity are the optimal conditions for making snow. But the weather can change from one place on the mountain to the next. So modern snow-making systems use computers to constantly monitor and adjust the machines.

First and 10, Virtually Speaking

In football's television broadcasts, an on-screen yellow line marks how far the team with the ball must go to make a first down. The simplicity of the line belies the technology that makes it happen. The trick is to make the line seem painted on the field and to have the players appear to run over it. What seems to be done with smoke and mirrors is actually an intricate process that includes rigging cameras and mapping the field.

Setting the Stage

MAPPING CONTOURS

A laser placed in the center of the field is used to collect data on elevation points. That information is used to draw a computer map of the contours of the field, and the map is overlaid onto the camera's view of the actual field.

THE VIRTUAL MEETS THE REAL

The computer map is tweaked and twisted so it matches the field as seen by each of the three main cameras as they pan, tilt and zoom. For each of the three camera views below, for example, the computer map image is lined up with the corresponding part of the camera image. All three cameras repeat this process at 10-yard intervals on both sides of the field.

Camera's view of corner of goal line

Far-side yard line

Near-side yard line

Camera's full field of view

FOLLOWING THE ZOOM

As cameras zoom in and out, the virtual line must follow suit, staying in perspective and getting larger or smaller.

To calibrate the system, the camera is set at full view and the computer map is adjusted to match. The calibration is repeated several times, as the camera zooms in, step by step.

Zoom

More zoom

TRACKING THE SHOTS

Computers need to know what the on-air camera is viewing to display the virtual yellow line accurately.

THE METHOD

❶ PAN ENCODER Measures the degree of movement when the camera pans from side to side.

❷ TILT ENCODER Measures, in degrees, up and down movements.

❸ REMOTE CAMERA SENSOR Collects the data on panning, tilting and zooming, encodes it and sends it to the system's computers.

CAMERA POSITIONS
Three cameras are used in the process. One is at the 50-yard line, and the others are at about the 25-yard lines.

THE LOOK OF THE LINE
The appearance of the line can be changed to look like paint on artificial surfaces or chalk on grass fields.

THE VIRTUAL FIELD
The computer-generated map of the field appears as a blue grid on the computers used. It is manipulated to fit the cameras' views.

Source: SporTVision

During the Game

AT THE HELM

Four people run the system on game day.

SPOTTER Inside the stadium with a view of the whole field. The position of each first down is relayed to the truck by radio.

LINE-POSITION TECHNICIAN Gets the first-down information from the spotter and enters it into a computer. Monitors the position of the line and makes adjustments.

1ST & TEN OPERATOR Monitors the broadcast line for quality and position. Uses another computer to check the colors of the field continually.

TROUBLE-SHOOTER Looks for problems and fixes them.

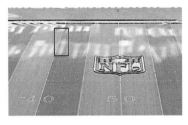

IN THE TRUCK
Scott Simon and Drory Yelin readying Sportvision's 1st & Ten computers before a National Football League game.

TROUBLE SPOTS
Field colors change constantly, depending on sunlight, clouds and stadium lighting. Computer data is updated so the line can be drawn even as shades change.

DRAWING THE LINE

SporTVision's system uses five computers to draw the virtual first down line.

GATHER PC Receives the pan, tilt and zoom data transmitted from the cameras.

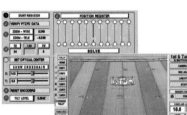

TALLY Keeps track of the on-air camera and tells the F Ten computer which computer map to use.

F TEN Displays the on-air video, overlaid with the right computer map. The line is drawn by specifying the yard line, to nearest tenth of a yard. For example: left, 37.8.

MATTE Tells the Render computer how to draw the line so the players will appear to run over the line.

RENDER Receives data from the other computers, then draws the line.

THE FINAL RESULT
All things lead to this: a virtual yellow line drawn at the first-down mark, in perspective, on top of the field and under the players.

Joe Ward and John Papasian;
Photographs by Bill Stover

When the Game's on the Line, The Line's on the Screen

By Matt Lake

Football, it's often been said, is a made-for-TV sport. For one thing, it can be so much more pleasant to watch the game from the comfort of a couch than from frigid seats high above the end zone.

Here's one more advantage that stay-at-home fans have: they can see exactly how far their team has to go to make a first down. Instead of squinting at linesmen, chains and poles, television viewers can see a computer-generated yellow stripe across the field, marking the line that the offensive team had to cross to make a first down.

This simple addition to football broadcasts, introduced on Sept. 27, 1998, in a game between the Baltimore Ravens and the Cincinnati Bengals, has been well received by viewers. Since that first broadcast, by ESPN, all the sports networks have adopted the technology.

Adding the line to a game broadcast is an impressive feat of logistics and engineering. It takes surveying equipment, special video cameras, a rack of computers and a staff of operators to insert a simple line and change it every time a first down is made. And it takes impressive computing power to generate and overlay the line.

The system receives raw video feeds from the three main cameras used in all broadcasts — at the 25-, 50- and 25-yard lines — as well as the video feed from the network's production unit that is about to go out over the airwaves.

When the offensive team makes a first down, a crew member on the field calls in the location of the marker and a computer operator in the video production outpost enters the measurement into the system. Graphics workstations generate a precise position and dimensions for the line, using information from a 3-D computer model and the live camera's positioning sensors.

Overlaying the line onto the broadcast image is the last stage of production, and it is no simple feat. The process is similar to chroma-keying — the trick used in weather reports where the broadcaster stands in front of a solid green or blue screen, and a weather map is superimposed electronically onto the screen.

Regular chroma-keying replaces a single key color with an overlaid image, but the grass or turf on a football field contains a range of colors, from shadowy dark greens to mud brown, and the colors can change as daylight fades or as clouds pass in front of the sun. A dedicated operator needs to tell the system to overlay a range of field colors — while excluding the colors of the teams' uniforms so the players will appear to step over the line.

Set-up Job

Even before a bowling ball plows through the pins with a satisfying thwack, microprocessors are choreographing what comes next: lifting and replacing the pins still in play, removing the rest and keeping track of the score.

1 BOWLER ROLLS THE BALL.

2 BALL TRIGGER

A small box near the pins contains a photoelectric eye, which transmits an infrared beam to a reflector on the opposite side of the lane. When the ball passes, the light is blocked briefly. The ball trigger uses the length of this period of darkness to compute the speed of the ball, and it alerts the pin spotter to begin cycling.

3 DIGITAL CAMERA

A digital camera (using a charge-coupled device, or C.C.D., chip) is located in front of the pins. One camera serves two lanes. The picture information is converted to pin fall data and then sent to the pin spotter and scoring computer.

NECK AREA

The camera takes a picture of the pins before and after the ball hits them. The first one is used to calibrate the system. The second, a comparison image, is taken after the ball is thrown. It shows if any pins are left standing, showing a vertical light stripe in the neck area of the image.

FRONT DESK COMPUTER
BOWLER'S TERMINAL
OVERHEAD TV MONITORS
SCORING COMPUTER
① APPROACHING BALL
LANE 2   LANE 1

BALL LIFT
AT THE BOWLER'S END OF THE BALL RETURN LANE
BALL
LIFTING WHEELS DRIVEN BY AN ELECTRIC MOTOR

② BALL TRIGGER
③ CAMERA
④ PIN SPOTTERS
DISTRIBUTOR PLACES PINS IN THE STORAGE AREA
MACHINE CONTROL COMPUTER
⑤ BALL RETURN LANE

ROLLING CARPET
ROTATING PIN WHEEL
BALL RETURN BELT BETWEEN A PAIR OF PIN SPOTTERS

Computers in a Bowling Center

The front desk computer has overall control of the system. It is used to activate lanes, compute charges, keep league statistics and create financial reports for the manager.

FRONT DESK COMPUTER	
SCORING COMPUTER	SCORING COMPUTER
BOWLER'S TERMINAL	BOWLER'S TERMINAL
PIN SPOTTER C.P.U.	PIN SPOTTER C.P.U.
CAMERA (PROCESSOR)	CAMERA (PROCESSOR)
LANE 1 LANE 2	LANE 3 LANE 4 ETC.

OVERHEAD TV MONITORS

The overhead monitors, run by the scoring computers, are used to display scores. Computer-generated graphics and animations can be used to indicate pins that were left standing and a suggested angle to pick up the spare.

4 PIN SPOTTER

The pin table descends and, using clamps, picks up the pins that are still standing, then replaces them. American Bowling Congress rules say the pins must be put back exactly where they were after the ball came through (a ball can sometimes cause a pin to move a couple of inches without knocking down the pin).

5 BALL RETURN

A special round door senses the ball's weight and lets the ball exit from the pin spotter pit. A rotating belt propels the ball into the return lane. Some ball returns use a ramp (and gravity) to return the ball.

PIN SPOTTER CYCLE

A One set of pins is ready on the pin deck, another set is in the pin storage area above. A rolling ball interrupts the ball trigger light beam and starts the cycle.

B The ball and knocked-down pins fall back into the pin spotter pit, a rolling carpet system. A pin table is lowered to pick up the standing pins.

C The table lifts up, and the sweep lowers and wipes the rest of the fallen pins into pit. A pin wheel brings them up to pin storage, which can hold two sets of pins.

D The remaining pins are reset and the table and sweep are brought up; pin spotter is ready for the second ball.

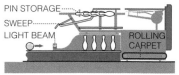

PIN STORAGE
SWEEP
LIGHT BEAM
ROLLING CARPET

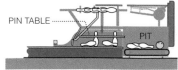

PIN TABLE
PIT

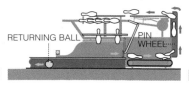

RETURNING BALL
PIN WHEEL

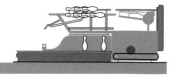

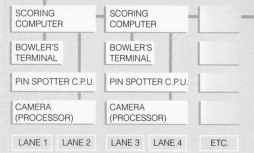

Source: AMF

Mika Gröndahl

Where All Lanes Are Fast Lanes: The High-Tech Bowling Alley

By David Kushner

Pity the pin boys. They were the shadowy tough guys at the ends of the bowling lanes who risked life and limb to line up the pins, sometimes for as little as a nickel a game. Extinct for decades, they are the dodo birds of bowling.

At high-tech lanes like Chelsea Piers in Manhattan, mechanics now tend digital cameras and the software of the electronic pin boys. Bowling, said to date from ancient Egypt, has grown increasingly high-tech in recent years as bowling equipment manufacturers have sought to transform it into a faster game for the digital age.

The sport has come a long way since 1946, when AMF Bowling introduced the first automatic pin spotter in Buffalo. Developed in an empty turkey house, the pin spotter used a mechanical suction system to lift the pins and set them back down. Newspapers marveled. "Robot Pin Boy Never Shows Fatigue," said a headline in The Brooklyn Eagle.

Today, the pin spotter is one part of a digitally networked system that bowlers interact with soon after they step through the door at most bowling alleys. When bowlers check in, clerks use a Windows NT Ethernet local area network to assign lanes and make changes if necessary.

Keeping score with pencil and paper is a thing of the past. Automatic scoring has become the norm. Players simply punch in their names at a scoring terminal and the system takes over, displaying and updating the information on monitors hanging overhead (an unfortunate innovation for bowlers having a particularly bad game). Computer animations appear between players' turns and display the best angles for the next shot. System crashes are rare.

A typical bowling cycle, which runs from the moment a ball is released until it is returned, lasts only about 8 seconds, according to AMF, which has its headquarters in Richmond, Va., and is one of the leading producers of bowling equipment around the world. That's half the time a cycle took 50 years ago. Speed has become a major selling point for competing manufacturers like AMF and the Brunswick Corporation of Muskegon, Mich.

To make lanes more consistent and quick in the new high-tech bowling alleys, pine and maple boards have been replaced by synthetic surfaces (a covering made from digital images makes the floor look like wood). The first two-thirds of the lane is lubricated with a viscous oil-based conditioner; the last third is kept dry to increase the ball's spin near the pins.

Just before striking the pins, the ball passes a ball trigger, an electric eye that tells the pin spotter to begin cycling in three seconds. That is the amount of time designated by the American Bowling Congress to insure that the pins have had enough opportunity to fall down and spin. After the ball strikes the pins, an infrared camera takes a digital image that tells the spotter which pins to lift and replace.

After the second ball (if necessary) is played, the 10 pins are replaced and the score updated on the screen. A printout of the game can be picked up at the front desk. The process is efficient, and that speed and ease translate into increased profits, said Randy Daniel, vice president for manufacturing at AMF.

Both AMF and Brunswick have injected some razzle-dazzle into their high-tech innovations. Brunswick came first with Cosmic Bowling, which features thumping dance music and fluorescent pins and balls. AMF followed suit with Xtreme bowling, its own version of what Mr. Daniels described as disco bowling; bowlers hurl glow-in-the-dark boogie balls, as they are called, as music plays. On either side of the lane, blinking red bumpers flash like airline runways. The bumpers keep every ball in play.

Is anyone nostalgic for the grittier days of pin boys and pencil-smudged scorecards? "Only the motion picture scouts," said Bill Lemon, owner of Bowlmor Lanes, a 60-year-old bowling alley in Greenwich Village, which moved to automatic scoring in the mid-1990's. "They love to find those old scoring desks for their films. Other than that, I don't think anyone misses them."

On the Open Ocean, on the Edge

Ocean sailboat racing requires a combination of skill, courage, a good technology that a sailor can afford. Here's a look at how one boat ends in France, covering 26,000 miles in between.

bit of luck and — particularly when it's a single-handed race — the best was equipped for the Vendée Globe, a nonstop solo race that starts and

The Boat

SAILS The sails consist of a mainsail, staysail and genoa for upwind sailing, and spinnakers (not shown) for downwind runs. The mainsail, whose winged shape and composite construction provides strength and efficiency, can be lowered — or reefed — in building winds. The staysail and genoa are on furling drums that allow them to be rolled up or unrolled in seconds.

MAST Supported by a rig of stainless-steel rods connected through three spreaders, the light, bendable carbon-fiber mast is the structure off which the sails are set. Sailors are occasionally forced to scale the mast in gale-force winds to make repairs.

MAINSAIL ·······

AUTOPILOT While the racer is sleeping or busy with other chores, a computerized steering system, with an electronic compass, automatically steers the boat for hours at a time.

RIGGING ·······

SATELLITE DISH ·······

RUDDERS ·····

CANTING KEEL To compensate for the forces of wind, a massive keel bulb underneath the boat acts as a counterweight by swinging outward to either side. This hydraulically controlled keel, which weighs several tons, can also right a capsized boat.

KEEL

Fore and aft water ballast tanks promote stability.

SPREADER

RADAR

NAVIGATION STATION

STAYSAIL

FURLED GENOA

60-FOOT CARBON FIBER HULL

KINGFISHER

888

Source: Merfyn Owen, Owen Clarke Design Group

Living Onboard

NAVIGATION A satellite-based positioning system is used. Outside routing assistance is prohibited, but sailors can download satellite weather images and access Internet weather sites. Also via satellite, they can transmit photos and video clips.

RESTING Navigating, forecasting, sail handling, maintenance and eating leave little time for sleep. Sailors often take catnaps throughout the day, and can average 5-6 hours of sleep each day.

ENDURANCE Nutrition and energy are supplied by a menu of weight-saving, freeze-dried food. An all-weather survival suit and layered undergarments keep sailors protected from the extreme weather.

DIRECTION OF WIND

A FASTER ROUTE

THE SHORTEST ROUTE

Southern Ocean Strategy

In the Southern Ocean, low-pressure systems spin clockwise around the center of the low. The skippers strive to use

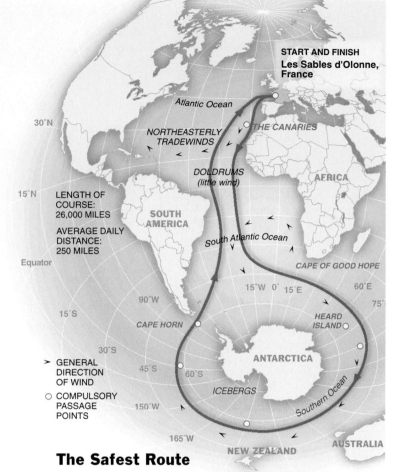

START AND FINISH
Les Sables d'Olonne, France

Atlantic Ocean

30°N

NORTHEASTERLY TRADEWINDS

THE CANARIES

DOLDRUMS (little wind)

AFRICA

15°N

LENGTH OF COURSE:
26,000 MILES

AVERAGE DAILY DISTANCE:
250 MILES

SOUTH AMERICA

South Atlantic Ocean

Equator

CAPE OF GOOD HOPE

15°W 0° 15°E

60°E

15°S

90°W

75°

CAPE HORN

HEARD ISLAND

30°S

> **GENERAL DIRECTION OF WIND**

○ **COMPULSORY PASSAGE POINTS**

45°S 60°S

ANTARCTICA

ICEBERGS

150°W

Southern Ocean

165°W

NEW ZEALAND

AUSTRALIA

The Safest Route

In the Southern Ocean, the stormy sea around the bottom of the planet, westerly winds flow unimpeded by land, and the boats can cover well over 400 miles a day in ideal conditions. Since the shortest distance to sail would take the racers close to Antarctica — and into iceberg territory — race organizers have established satellite-monitored passage points to corral racers north and out of harm's way.

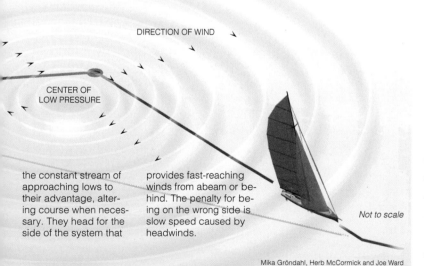

DIRECTION OF WIND

CENTER OF LOW PRESSURE

the constant stream of approaching lows to their advantage, altering course when necessary. They head for the side of the system that provides fast-reaching winds from abeam or behind. The penalty for being on the wrong side is slow speed caused by headwinds.

Not to scale

Mika Gröndahl, Herb McCormick and Joe Ward

So Easy, Even a Weekend Sailor Can Do It

By Catherine Greenman

Around-the-world sailors like those who take part in the Vendée Globe race depend on advanced technology to finish — or at least to survive. But even more casual boaters are making use of high technology, not the least of which is electronic navigation equipment and software.

Instead of plotting a course with compasses and paper charts, for example, a sailor can use a Global Positioning System receiver to determine the boat's latitude and longitude, and then plug that position into navigation software on a laptop computer. The software can tell the boater which way to head and track the journey on digitized nautical charts on the computer screen. It can even be hooked into an autopilot that will steer the boat.

The software can also be used to plan trips — an around-the-world one for a Vendée Globe racer, or a more modest trip along the New England coast for the weekend sailor. A boater sailing from New York to Rhode Island for the weekend, for example, could plan various routes by clicking on a series of destination points on maps of the Long Island and Block Island Sounds. The software calculates the distances between destination points, total distance and estimated travel time. It's easier than figuring out a route by hand.

With a route planned, the boater plugs the G.P.S. device into the computer and lets the software point the way. Most navigation programs show a trail of "bread crumbs" on the chart, providing a record of the route.

Although navigational aids like G.P.S. have made it easier for less experienced boaters to pilot sailboats and powerboats with more confidence, experienced sailors, sailing organizations, and even the manufacturers of navigational software agree on the dangers of overreliance on the technology.

"Electronic navigation has become so reliable that we joke that you could program your computer and GPS to sail your boat out of the harbor in South Carolina and take a plane to meet it in Florida," said Milt Baker, president of Bluewater Books and Charts, a company based in Fort Lauderdale, Fla., that sells electronic charts. "But one stray electron or one dead battery and you're out of business. You have to know how to navigate and read charts on your own."

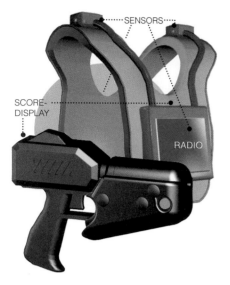

VEST Made of polyethylene foam, the eight-pound vest contains four sensors (one on each shoulder, the chest and the back) that register hits in all directions from the phasers' infrared rays. When hit, the sensors vibrate and flash. The vest has a digital readout on the front displaying how many shots and lives the player has remaining, plus the score and the vest's identification number.

PHASER This plastic, soft foam-insulated phaser is connected by a cord to the vest. When fired, each phaser produces two beams: laser and infrared. The laser beam is harmless. It's the invisible infrared ray that transmits all the necessary data to sensors on the vests and bases.

Armed, but Not Dangerous

Here's a schematic diagram of how a laser tag arena can be set up.

1 WHEN THE GAME STARTS After obtaining game-play data from the game computer, the transceiver signals the vest packs to update to the appropriate number of shots and lives. Once the vest packs are charged up, the game begins, with players shooting at the opposing team and its base sensors. Every second, the transceiver polls the vests for new information by sending a radio signal to and from each vest. Along the way, the information is translated into digital information for the game computer.

2 SCORING Each player's infrared ray carries an identifying piece of data. Say Red Player 8 hits Green Player 7's sensor. When the transceiver polls the players' vests, it will be alerted that Red Player 8 has hit Green Player 7. It will then alter their scores.

3 HITTING A BASE The scoring process is the same, except that the base sensors are linked directly to the game computer.

GAME COMPUTER A PC runs laser tag with customized software.

BASE RECHARGER

BASE STATION
Each team has a base station, fitted with sensors similar to those on the vest; a player can earn points by hitting one of the opposing team's base sensors. At each station, a recharger transmits radio waves and automatically updates players' vest packs with shots and lives each time someone passes it.

TRANSCEIVER The eyes and ears of the game, this one-square-foot box hangs from the ceiling and is hard-wired to the game computer. It converts the computer's digital data into radio waves, then transmits to the vest packs. Data coming from the vests are picked up by the transceiver, converted back to digital information, then sent back to the game computer.

SCORECARD At the end of the game, the computer kicks out a scorecard for each player with game statistics.

BASE SIREN LIGHT

BASE SENSOR

BASE RECHARGER

BASE STATION

INFRARED

LASER

RADIO SIGNAL

Sources: LaserTron; Actual Reality International; Lazer Park in Manhattan.

Mika Gröndahl

Laser Tag: An Old Child's Game Gets an Upgrade

By David Kushner

The fighter had just sent Sandman cowering through the fog when he heard the sound of his shoulder exploding. Sweeping his phaser around the pylons, he fired wildly into the darkness. Surely his nemesis could not have come back already; the enemy was probably at his recharger by now, loading up again on shots.

Sucker! The thought flashed through his mind as he was hit again, and his shoulder sensor pulsed with blood-colored light. Looking down — way down — behind a green neon-tipped pylon, he saw Sandman's teammate, Caster Troy. Alas, this was his lethal assailant — four feet tall, 9 years old, barely able to wrap her hands around her phaser.

So it goes in laser tag, perhaps the only sport besides jacks that allows a fourth grader to whip a 29-year-old weighing in at 200 pounds. And one of the things that levels the playing field is the technology, a sophisticated system of data exchange that works, quite literally, at the speed of light.

Inspired by "Star Wars," a Texas inventor named George Carter is said to have created the first laser tag game in the early 1980's. Mr. Carter built a string of Photon Laser Centers in Dallas. There are now hundreds of laser tag arenas, from Tampa, Fla., to Sydney, Australia.

Though the technology has improved, the essential concept remains the same. Like a science fiction version of capture-the-flag, laser tag is played in arenas from 2,000 square feet to 12,000 square feet. Players, either individuals or teams, don vests stitched with sensors; their weapons are so-called phasers that shoot infrared rays the sensors pick up. The phasers also fire red lasers, which are just for show. There's no danger — the lasers used for recreation must be approved by the Food and Drug Administration.

Lazer Park, just off Times Square in New York, is a typical laser tag arena, with an atmosphere that is right out of a science-fiction thriller. The arena features a maze of towering black plastic pylons, machines that pump water-based fog, and glowing black-velvet posters of lunar landscapes on the walls. A high-decibel soundtrack helps keeps the adrenaline flowing.

While the players are chasing and shooting their phasers at one another, the action is indeed fast-paced, although for safety reasons running is not allowed. Referees roam the labyrinths to catch runners and to make sure people are the proper distance away from each other, and penalize violators by suspending them from play. After repeated violations, some referees will hold a player's vest in the air and encourage others to riddle it with shots.

To be an accomplished laser-tagger requires lightning-fast reflexes, but not necessarily impeccable aim. The width of the infrared beam allows someone to score a hit by firing within several inches of a sensor. At the end of each game, players leave with a detailed printout of whom they hit and how often, and who hit them.

Many in the business would like to make laser tag a legitimate sport. Leagues compete weekly. Nearly half the players are teen-aged males (no surprise there). But increasingly adults are joining the fray. Eric Guthrie, executive director of the International Laser Tag Association in Indianapolis, said more and more players are coming from corporations, which use the events as team-building exercises.

At Lazer Park in Manhattan, the owner, Marc Epstein, credited the increasingly lightweight and simple game gear with the boom in business. "We try to provide something high-tech that will keep people coming back," he said. "If you have the right mix, you forget about your hard day at work and feel like you've been in another world. We put you in an alternate reality."

Home versions are also popular, but jumping over couches in one's living room just doesn't have the same oomph as dodging shots in an arena.

At Lazer Park, a 31-year-old player who would give only his nickname, Slippery Weasel, said he preferred the excitement of laser tag over PC shoot-'em-up games like Doom and Quake. After a particularly rigorous match, Slippery Weasel stood outside the arena dripping with sweat. "Instead of watching a video game," he said, "you're in it."

All Bets Are Online

When horses race at Aqueduct, Belmont or Saratoga race tracks in New York State, Thoroughbred fans can place bets at the tracks, at a variety of off-track betting parlors or by phone. Bets at the tracks are made through clerks or automated teller machines. This example shows how bets are handled for a race at Aqueduct.

Associated Press

Ticket-issuing machine, or TIM. ▲

Betting Network

The betting information for an Aqueduct race from an OTB parlor can be linked directly to the Aqueduct track through a dedicated phone line or dial-up modem or can be connected through an intermediary hub, like one at Saratoga or Belmont. The head office of the New York City Offtrack Betting Corporation serves as the hub for betting parlors in its area.

TOTE COMPUTER SYSTEM

MASTER ⋯ SECONDARY

CLONE

Craig Wallace Dale (above and top left)

OFF-TRACK BETTING PARLOR

(More than 200 in New York State.)

DEDICATED LINE AND DIAL-UP PHONE LINE FOR DATA TRANSMISSION

REGIONAL OFF-TRACK BETTING CORPORTATION HUB

(New York City OTB, with its head office in Manhattan, is one of the six regional OTB corporations.)

DEDICATED LINE AND DIAL-UP PHONE LINE FOR DATA TRANSMISSION

AQUEDUCT RACE TRACK

Aqueduct links to the hubs or directly to OTB parlors through dedicated lines and dial-up modems. Before a race, Aqueduct updates the hubs with any last-minute changes, like scratches.

OTHER HUBS

Before a race, the pertinent data are loaded into the hubs' computers: horses' names, time and date of race, odds, etc. That information is made available for gamblers. After betting closes, each hub collects betting data from the OTB parlors that are running in its system. Aqueduct can exchange data with dozens of in-state and out-of-state hubs. Those hubs can be at out-of-state tracks, like Santa Anita in California, or can, for example, be connecting casinos in Las Vegas.

The brain of the betting process at Aqueduct (or other hubs) is the **TOTE COMPUTER SYSTEM**.

1 Many times each second, the master tote computer polls each betting window at the track to get wagering details: the amount and type of each bet on each horse in a race. It simultaneously polls each hub and direct OTB link. It computes the odds for each race, based in part on the betting.

2 The master tote computer sends a copy of this data to the secondary tote computer, which confirms each bet and instructs the terminal or hub to issue an appropriate ticket.

3 The secondary computer then sends a copy of this data to the clone, or backup, computer. After the race is completed, the tote system furnishes the information for paying off each winning bet.

SIMULCASTS provide live audio and video satellite feeds from the race track to OTB parlors and hubs.

Source: The New York Racing Association Inc.

Mika Gröndahl

Racing's Brains: Handling the Action at the Tracks and Elsewhere

By David Kushner

And they're off!

It was Tomlin's Flag by four lengths as the ponies rounded the turn of the Aqueduct horse racing track in New York. The few thousand gamblers at the track weren't the only ones anxiously watching the dust fly. Thanks to a specially designed computer network, Thoroughbred fans across the country could be focused on the very same race.

On a typical Saturday afternoon, for every dollar that Aqueduct pulls in from gamblers at the track, it gets about $6 from bettors elsewhere around the state and nation. With a computerized betting network like the one employed by the New York Racing Association, betting from afar is easy. "There is an explosion in interstate wagering," said Bill Nader, director of broadcast communications for the group.

Off-track betting in New York began in the mid-1970's. Bets were recorded then on magnetic tape, and information was manually entered into Aqueduct's database; OTB's relied solely on radio broadcasts for updated information. Around that time, the racing association's tracks switched from an electromechanical wagering system, which had to be updated manually, to an electronic one called a Totalizator. This high-powered computer, nicknamed Tote, can process several hundred transactions per second.

At Aqueduct, the Tote resides in the windowless basement of the clubhouse. During a race, the Tote monitors the wagers at Aqueduct as well as at those placed at off-track betting parlors linked nationwide.

At the track, gamblers have three choices for placing a bet. Throughout the clubhouse are ticket-issuing machines, called TIM's, with touch screens. They are somewhat similar to an automated teller machine at a bank; a gambler simply punches in the type of bet, the horse and the amount of the wager, then receives a ticket recording the bet. If the gambler is lucky, that ticket can be redeemed for cash after the race. Bettors can open accounts and use bar-coded cards and PIN's to get access to the accounts.

Smaller ticketing machines, called Tiny TIM's, are in the clubhouse restaurant. And for those who would rather deal with people, clerks are ready to dole out tickets.

The Tote system continually polls the wagering stations at Aqueduct and elsewhere. Rather than running the data from hundreds of off-track parlors through one machine, data traffic is managed by 65 hubs.

Gamblers who have personal accounts with the racing association can place bets over the telephone. They can still see the races because races are simulcast on cable television in the city and elsewhere. And live audio feeds of the races are available through the New York Racing Association Web site. The site also contains information on horses' past performances, so a bettor who wants to see if a particular horse has ever finished in the money can, for a small fee, do the research online.

In short, a horse player can open an account with the racing association, get race information from the association's Web site, place a bet over the phone, then listen to the race on the Net. It's not Internet gambling, but it is gambling that makes use of the Internet for information purposes. Bets are placed over the phone.

Bumper to Bumper

Many of the visible features of pinball machines — electromechanical devices like bumpers and flippers — have changed little over the years. But behind the scenes (or rather, in back of the glass), even rudimentary pinball games have gotten a technological upgrade. Microprocessors and other chips and circuit boards monitor and control the action and display the score to the player.

THE BRAIN BEHIND THE GAME

The central processing unit, or CPU, is the brain of a typical pinball machine. The hardware is usually located in the back of the machine, behind the upright glass that faces the player. The software for each game is loaded into the CPU, which dictates every detail of the action: the rules, the timing sequences, the sounds, the music, the speech and the scoring. The input/output, or I/O, board is also behind the glass. Acting like a middle manager, it is connected to the CPU through a wire harness and ribbon cables, then to all the specific components of the playing field. Commands are issued from the CPU to the I/O board, which then signals the various components involved.

ON YOUR MARK . . .

When a coin is inserted, it trips a switch that is connected to a wire linked to the CPU. The CPU realizes that it needs, for example, another coin before play begins, so it flashes a half-credit message on the dot matrix display. After another coin is inserted, the CPU tells the I/O board to flash the lights near the Start button.

GET SET . . .

When the Start button is pressed, the CPU instructs the I/O board to reset all the target banks and playing field elements. The ball is then delivered through the shooter lane.

GO!

The player pulls a spring-activated plunger to send the metal ball into the playing field. Some games use an electronic trigger called an autolaunch.

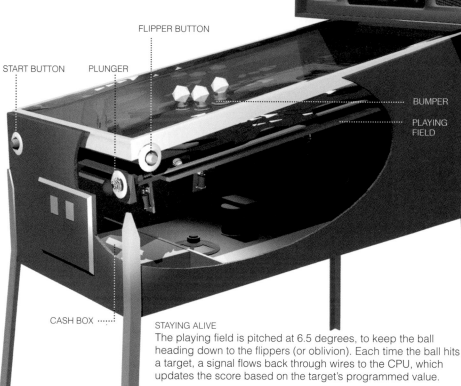

CPU
I/O BOARD
SPEAKERS
DISPLAY
13.074

FLIPPER BUTTON
START BUTTON
PLUNGER
BUMPER
PLAYING FIELD
CASH BOX

STAYING ALIVE
The playing field is pitched at 6.5 degrees, to keep the ball heading down to the flippers (or oblivion). Each time the ball hits a target, a signal flows back through wires to the CPU, which updates the score based on the target's programmed value.

TILT MECHANISM

Like any smart machine, a pinball game has to protect itself. Since nudging the cabinet has long been a necessary part of the game, Harry Williams, a pinball entrepreneur, invented a plumb-bob mechanism more commonly known as the "tilt."
A thin wire with a weight at the end simply hangs inside a metal ring. When a player gets too aggressive, the wire touches the ring, completing a circuit that sends a signal to the CPU. The CPU then instructs the game to turn off its lights and temporarily shut down.

RING
PLUMB BOB

BUMPER MECHANISM

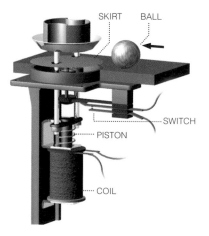

SKIRT   BALL
SWITCH
PISTON
COIL

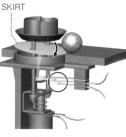

SKIRT

A BUMP
When the ball hits the bumper skirt, it tilts and bends a spoon-switch actuator with its rod. As the spoon bends, it completes the circuit in the switch and activates the coil.

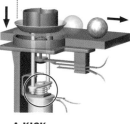

METAL RING

A KICK
The activated coil pulls down the metal piston, which is connected to the bumper's top ring. As the ring comes down sharply, it kicks the ball away.

Source: Sega Pinball

Mika Gröndahl

Pinball Machines: Give Me That Old-Time Technology

By David Kushner

Video might have killed the radio star, but what about the pinball manufacturer? Ever since Pong entered the arcades and bars in the 1970's, skeptics have been predicting pinball's last big tilt.

Faced with the declining popularity of their games in the late 1990's, the industry's two main manufacturers, Williams Electronics Machines and Stern Pinball, took two different technological paths, with one embracing and the other rejecting the video-game age.

After nearly six decades in the business, Williams Electronics, based in Chicago, announced in the late 1990's that it would no longer produce traditional pinball machines and instead would concentrate pinball-video game hybrids. The company's first such game, Revenge From Mars, was introduced in London in 1999.

While video had been used for a while to dress up pinball games, Revenge From Mars was the first to incorporate video images into the action, allowing the player to interact with virtual, as well as physical, targets.

But for Williams Electronics, even video glitz wasn't enough. Its parent company, WMS Industries, announced late in 1999 that it would no longer make any kind of pinball machine. The last machine, a Star Wars pinball-video hybrid, rolled off the assembly line in November of that year.

Stern Pinball, however, has remained committed to the classic electromechanical action of bumpers and wire ramps.

Williams's strategy was flawed, said Gary Stern, president of Stern Pinball. Williams, he said, was trying to get a video game player to play pinball, "but the video game player is at home playing Dreamcast" or other game console system.

Pinball has been around for 128 years, but this debate did not begin until video games took off in the early 1980's. In the beginning, pinball was based only on gravity. It originated in Cincinnati as a tabletop version of a French game, bagatelle. The player shot a ball into a slanted play field, where the ball would randomly bounce among a series of pins (thus the name).

In the 1930's, Harry Williams — the future founder of Williams Electronics — invented a game called Contact, which included the first battery-operated feature: a scoring cup that automatically kicked out a ball that landed in it.

As a joke, Mr. Williams connected the cup to a telephone so the phone rang every time the ball went in the cup. Legend has it that his unsuspecting business associate spent much of the day sprinting across the room to answer the phone each time it rang.

The flipper was introduced in 1946. The next major technological innovation came in the mid-1970's, when pinball's electromechanical elements were augmented with electronics. That allowed machines to have more dazzling features, like digital scoring, speech and synchronized light shows.

The first threat from video games came around the early 1980's, during the explosive era of arcade hits like Pac-Man and Space Invaders. Faced with this formidable competition, pinball developers experimented with hybrid designs in games like Caveman and Baby Pac-Man.

These machines embedded video game displays within the game so a player would switch, rather awkwardly, between pinball and video maneuvers. By all accounts, the early hybrids were a failure.

The newer hybrids were different, because the video elements were seamlessly integrated into the pinball action. When the ball passes various sensors on the playing field, corresponding video images are projected on the glass. In Revenge From Mars, for example, a player may have to shoot video images of six Martians who sit at a happy-hour bar.

Mr. Stern said that such flashy techniques are too much of a good thing. "Hybrids tend to be dismal because the average player gets confused by all the action," he said. The allure of pinball is its simplicity, he said, because players, most of them men 18 to 35 years old, come to the games for fast and intuitive play. "We don't want our players to think too much," Mr. Stern said.

Hit Factory

Music kiosks let consumers make compilations of their favorite songs on CD. Each kiosk has a touch screen, hard drives to hold a database of songs, a CD recorder, printers for labeling the disc and preparing the jewel-case insert and a computer to coordinate everything. Musicmaker.com's kiosk, shown here, takes care of the entire process at one spot. With other manufacturers' models, recording and printing take place elsewhere in the stores.

THE TOUCH SCREEN

Customers make selections from a list of music genres or artists, then from lists of songs. They are able to listen to samples of songs before making up their minds. Selected songs can be arranged in any order. Customers can type in names for their discs and select clip art for the labels.

BILL VALIDATOR AND CARD SWIPER

The kiosk accepts cash or credit payments. The bill validator and credit card reader are standard devices used in things like vending machines and A.T.M.'s.

COMPUTER

A dual Pentium chip machine runs the process.

HARD DRIVES

The kiosk has five hard drives with a total capacity of close to 200 gigabytes — enough space to store about 5,000 songs in the format used for standard audio CD's.

PRINTERS

A thermal printer prints the name of the disc, the song list and any art selected on the top side of the disc. Another printer in the top section prints a paper insert for the CD's jewel case.

CD RECORDER

An 8X recorder can record a 32-minute CD in about 5 minutes, including setup time. It uses CD-R discs, which can be recorded on only once.

THE FINISHED PRODUCT

The CD comes out of a slot on the right side of the kiosk; the jewel case and a paper insert for it come out of separate slots. The machine holds about 100 cases.

SAMPLING SELECTIONS

Speakers and headphones let the user hear snippets of songs before selecting them for the CD.

musicmaker.com

JEWEL BOX DISPENSER

CD EXIT SLOT

BLANK CDs

Sources: Musicmaker.com; Olea Exhibits/Displays Inc.

Frank O'Connell

Appearing at a Store Near You: An A.T.M. for the Ears

By Karen J. Bannan

Sharon Gordin loves music. She is always on the lookout for the latest dance compilation or trendy album. She is also enthralled by the idea of downloading MP3 music files onto her computer for her own use.

But Ms. Gordin, an executive assistant from New York, felt left out of the MP3 craze. She didn't have a broadband Internet connection to make quick work of downloading music files, or a CD recorder to create her own music CD's. And anyway, she said, it all seemed rather complicated.

For Ms. Gordin, music kiosks may be the answer. They let people create their own discs in a way that is almost as convenient as buying a prepackaged one.

There are several different types of music kiosks. But they are all designed with the same fundamental goal: to let consumers select songs and create music CD's right then and there.

"Kiosks give a similar experience to MP3 files," said Pierre Tager, vice president for engineering at Musicmaker.com, one kiosk manufacturer. "Customers can browse thousands of tracks, personalize their CD and take it home."

Kiosks fill a need, as shown by the success of MP3 files, said Mark Mooradian, an analyst with Jupiter Communications, a research firm. "We've seen an explosion of compilation discs," Mr. Mooradian said. "People normally don't want an entire album anymore. That's one of the reasons MP3's are so popular."

Musicmaker.com, based in Reston, Va., and Liquid Audio, in Redwood City, Calif., make kiosks that let consumers create their own compilation discs, with thousands of singles to choose from.

But Digital On-Demand, in Carlsbad, Calif., offers only entire albums. It has relationships with most major record distributors and offers up to 5,000 albums for purchase. Most of these albums are selections that the store wouldn't normally carry, said Tom Szabo, chief executive officer of Digital On-Demand.

Each of the manufacturers' kiosks has a touch screen so users can browse through a library for songs or albums. Like their ancestor, the listening kiosk, the music kiosks can be used by consumers who want to hear 30-second samples of music. Liquid Audio uses a secure private network to update its kiosks with new songs, but the kiosks can also be connected to a song database on the Internet.

Once one or more songs are selected, the digital music files are recorded onto a blank CD. Although MP3's have become the de facto music standard on the Internet, kiosk providers use various file formats to burn CD's. Users can also select jewel-box album covers and clip art for a customized cover.

Musicmaker.com's kiosk is self-contained, with a CD recorder and printer, and a credit card reader for payment. Digital On-Demand and Liquid Audio send digital files across an in-store network to printing and recording equipment behind the sales counter.

Burning a CD isn't quick. Musicmaker.com's kiosk has an 8X CD-ROM burner within its cabinet. If an album is 32 minutes long, it takes 4 minutes to burn it onto a disc. Add that 4 minutes to the time it takes for a consumer to browse, select songs and pay for the choices, and it is apparent that the number of people who can use the kiosks in an hour is limited. Digital On-Demand's kiosk could take just as long, depending on the number of CD burners behind the counter.

But CD's are only the first step, Mr. Mooradian said. The same kiosk that burns CD's could also download music files into portable MP3 players (for a fee, of course). And if songs can be downloaded, it may not be long before kiosks have DVD movies, software and electronic books on tap as well.

Data In, Music Out

MIDI, or Musical Instrument Digital Interface, is a standard format that allows multiple synthesizers to be linked together through the use of controllers — computers, piano-type keyboards or other MIDI instruments. With MIDI, a person playing a controller that looks like a wind instrument can instruct a synthesizer to create a completely different sound — that of a drum, for instance. MIDI includes a common language that provides information about music, like note on or off, sustain, pitch bend and timing. But MIDI files contain only instructions for creating music, not the sound itself.

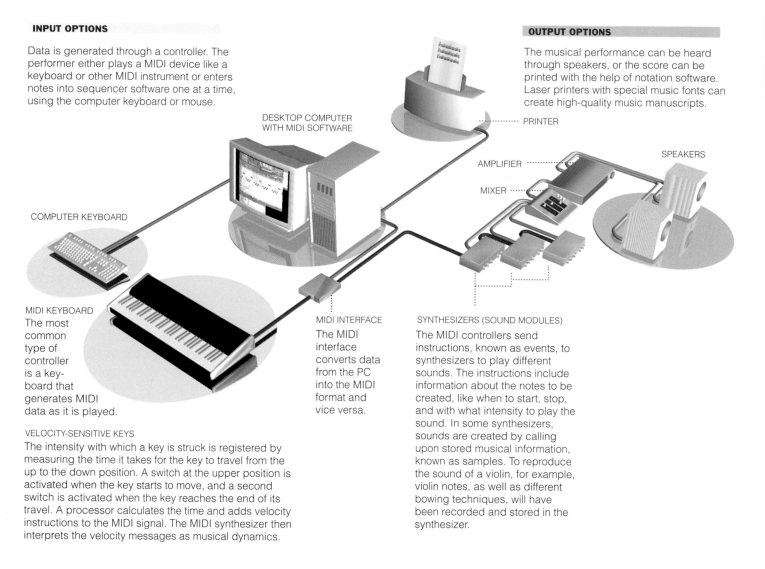

Data is generated through a controller. The performer either plays a MIDI device like a keyboard or other MIDI instrument or enters notes into sequencer software one at a time, using the computer keyboard or mouse.

The musical performance can be heard through speakers, or the score can be printed with the help of notation software. Laser printers with special music fonts can create high-quality music manuscripts.

DESKTOP COMPUTER
WITH MIDI SOFTWARE

.... PRINTER

SPEAKERS

AMPLIFIER ....

MIXER ....

COMPUTER KEYBOARD

MIDI KEYBOARD
The most common type of controller is a keyboard that generates MIDI data as it is played.

MIDI INTERFACE
The MIDI interface converts data from the PC into the MIDI format and vice versa.

SYNTHESIZERS (SOUND MODULES)
The MIDI controllers send instructions, known as events, to synthesizers to play different sounds. The instructions include information about the notes to be created, like when to start, stop, and with what intensity to play the sound. In some synthesizers, sounds are created by calling upon stored musical information, known as samples. To reproduce the sound of a violin, for example, violin notes, as well as different bowing techniques, will have been recorded and stored in the synthesizer.

VELOCITY-SENSITIVE KEYS
The intensity with which a key is struck is registered by measuring the time it takes for the key to travel from the up to the down position. A switch at the upper position is activated when the key starts to move, and a second switch is activated when the key reaches the end of its travel. A processor calculates the time and adds velocity instructions to the MIDI signal. The MIDI synthesizer then interprets the velocity messages as musical dynamics.

SOFTWARE
Sequencing software is the part of the MIDI chain that records musical information and issues instructions for the playing of MIDI data. Using the software, composers can quickly and easily alter the pitch, volume and instrumentation of their composition.

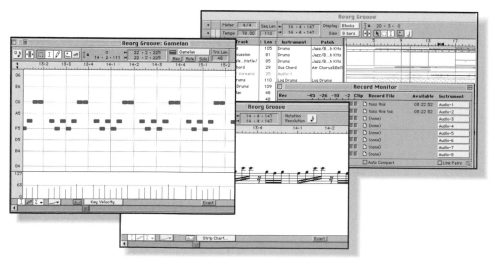

Mika Gröndahl

MIDI Lets Electronic Devices Make Beautiful Music Together

By Eric A. Taub

The music industry seems to continually battle over which technology should become the standard for listeners who want to download small, secure, good-quality audio files from the Internet.

Composers, on the other hand, have had a standard workable digital form for years. Musical Instrument Digital Interface, known as MIDI, gives composers, using a computer and special software, the ability to simulate the sound of an entire symphony orchestra or quickly change a composition's key, instrumentation and tempo and hear the results instantaneously. And it doesn't require a single performing musician.

MIDI is an industry-standard protocol that lets composers easily connect a computer with a keyboard and musical synthesizers. Before MIDI was unveiled in 1983 by a group of musical instrument manufacturers, electronic musical instruments were connected to each other by different methods. This often made it impossible to use two instruments from different manufacturers.

Now, to compose and edit music on a PC, a musician purchases a low-cost MIDI interface box, which connects to a computer via the serial port, or installs a MIDI card. Music sequencing software is used to record the notes, as well as the composition's instrumentation, key and length.

To play music in real time, a controller, which can be a MIDI-compatible keyboard or wind or string instrument, is added. The setup will also include at least one synthesizer, which actually creates the sounds. Some synthesizers manipulate samples of different instruments that have been stored in its memory. Those sounds are then fed to an amplifier and speakers to become audible.

With this MIDI-compliant package, a composer today can create compositions that to most listeners sound as if they are actually being played by scores of musicians seated in a concert hall.

"MIDI gives composers the musical equivalent of a word processor," said Thomas Dolby Robertson, founder of Beatnik, which produces technology to deliver high-quality music and sound over the Internet.

A MIDI-compatible computer can control up to 16 different channels on each of an infinite number of synthesizers that can be chained together, giving composers the ability to create music with innumerable layers of sounds.

Unlike MP3 or the other musical compression formats, MIDI files are not composed of musical sounds; rather, they are digital instructions telling the synthesizers what notes to create, how long to play them and with what intensity.

MIDI's small file size makes it ideal for CD-ROM games, in which space is at a premium. A game player hears music because the disk's embedded MIDI data instructs the PC's sound card to create certain notes. But those notes are not embedded in the game itself.

Composers, equipped with very expensive synthesizers, use MIDI to create compositions that would be impossible or extraordinarily expensive to perform with live musicians.

Jeff Rona, who has composed musical scores for the movie "White Squall" and the television shows "Chicago Hope" and "Profiler," uses MIDI and computer sequencing software to create compositions performed by virtual orchestras. Changing the sound from a violin to a cello or tuba is a simple matter of the PC instructing the synthesizer to use a different instrument.

Mr. Robertson samples unusual sounds, like crickets and trees falling in the forest, and then using MIDI commands instructs the synthesizer to substitute those for cymbals and bass drums.

Because of the complications of human speech, the MIDI standard will remain confined to controlling tones, like instrument sounds and sound effects, that can be accurately sampled and then reproduced electronically by synthesizers. "Voice sampling is rarely done," said Tom White, president of the MIDI Manufacturers' Association. "Speech is extremely complicated. There are so many nuances, mouth shapes and tones that it's very difficult to create virtual speech from a recorded sample."

Some MIDI software has a "humanizer" element that subtly alters tempo and dynamics so that the music does not sound machine perfect. But can music created using synthesizers and sampled tones, and controlled by MIDI devices, ever sound exactly like the real thing?

"MIDI-created music can't, because physical sounds are intrinsically linked to underlying physical gestures," said David Mash, vice president of information technology at the Berklee College of Music in Boston. "If you play a violin sound on a keyboard, you can't play the keyboard as if it's a violin. "When you're one person, trying to perform 15 different parts of a composition by yourself, you've got to try and inject those missing personalities that would be there in a live performance. It's very difficult to do so. Still, it's a lot cheaper than hiring a whole symphony orchestra."

Fine-Tuning a Hall

Selected characteristics of the Tokyo Opera City concert hall illustrate how daring architecture can mesh with precise acoustics.

DIFFUSION

If a hall diffuses sound properly, it softens the sound field as it is reflected off different surfaces.

LARGE-SCALE DIFFUSION

Tests revealed that the reflected sound was best when balcony fronts were sloped forward and stepped. The curved bottom edges help distribute the sound uniformly to the audience.

FINE-SCALE DIFFUSION

Small irregularities on the lower side and rear walls scatter high-frequency sound and reduce "acoustical glare" that can occur with smooth surfaces.

Sound-reflecting blocks scatter the sound laterally where it strikes other parts of the ceiling to further diffuse the sound.

BALCONY
FRONTS

20 – 24
DEGREES

The rear sloping wall is covered with diffusers to eliminate echo.

STAGE

66 Ft

The canopy distributes sound back to performers and to front portions of the audience.

INTIMACY

The closeness that an audience feels toward performers can be attributed to a short gap in time between the initial sound and the arrival of the first reflected sound. The hall's 66-foot width keeps this gap within an optimum range.

SPACIOUSNESS

The sense of being enveloped by sound reverberations is aided by the open space above the balconies in combination with the narrowness of the lower hall, where sound can reflect laterally toward listeners.

BASS RATIO

It is a measure of the strength of reflected bass tones. Since the wood surfaces absorb more bass than plaster, seats were chosen that absorbed less bass, and no carpets or other absorbent materials were used.

Sources: Acoustical Society of America; Dr. Leo L. Beranek; Takayuki Hidaka, Takenaka R & D Institute

Steve Duenes and John Papasian;
Photographs courtesy of Tokyo Opera City Arts Company

Acoustics: Finding Aural Warmth in Cold Hard Silicon

By James Glanz

Can the sense of acoustic intimacy created by a fine concert hall be measured in how many milliseconds it takes sound waves to ricochet from the walls and balconies and reach a listener in the seats? Can a hall's aural warmth be calculated from how efficiently bass notes rebound from the same surfaces? Can the prized quality called resonance be estimated from the rate at which the entire hall fades to silence after a blast of electronic sound?

More to the point, can an architect rely on studies of these quantities, using computer calculations and measurements in scale models, to ensure that a new concert hall will be an acoustical triumph rather than a disaster?

For years, the answer to all these questions seemed to be no — the field of concert hall acoustics has had only spotty success. But an unusually intense collaboration between architects and acousticians put the science of acoustics to the test, with two major successes in Tokyo.

The halls in question are the 1,632-seat concert hall of the multipurpose complex called Tokyo Opera City, and the 1,810-seat opera house of the adjacent New National Theater. Both have architecturally daring designs, yet both have been praised by musicians who have performed in them.

"This hall simply has some of the best acoustics in which I have ever had the privilege to play," the cellist Yo-Yo Ma wrote in a commentary on the concert hall that appeared in a technical journal.

The research for the new halls, whose principal architect was Takahiko Yanagisawa, president of TAK Architects in Tokyo, may be the most extensive use yet of acoustical measurements and calculations in efforts to design concert halls that are not simply copies of great halls of the past.

"If you make a copy of the old, great halls, you'll have a great hall," said Dr. Leo L. Beranek, an architectural acoustician in Cambridge, Mass., who was the principal acoustical consultant for the projects in Tokyo. But the Tokyo concert halls, he said, "are different in appearance and they have the sound of great halls."

As the designs of the halls took shape, scientists analyzed and worked to maintain acoustical variables like reverberation time, spaciousness and intimacy, each with a precise mathematical definition and musical meaning.

Without those studies, "you're gambling" on the acoustics, Dr. Beranek said.

The researchers also used additional acoustical measurements, or metrics, developed by Dr. Beranek, to quantify the sound. Each of these metrics has a precise mathematical meaning that seeks to isolate a specific aspect of acoustical quality in a hall. In the studies leading up to the design of the new Tokyo halls, measurements of those and other metrics were made in 20 opera houses and 25 symphony halls in 14 different countries.

The idea was to get a quantitative measure of what made the good halls good and the bad ones bad. For the studies, the researchers generally produced a burst of sound from a 12-sided speaker on the stage. Each burst and its acoustic aftermath was recorded on tiny microphones placed in the ears of dummies, and in some cases the ears of real people, scattered around the seats. The team worked out the value of the various metrics for each hall by analyzing detailed forms of the sound waves picked up by the microphones.

Dr. Beranek and his collaborators then compared those measurements with an acoustic ranking of the halls based on a survey of conductors and music critics. They found that the most beloved concert halls had reverberation times, for example, near two seconds.

Then the acousticians turned to large computers that had been programmed to simulate the acoustics in the basic architectural designs of Mr. Yanagisawa.

The team eventually built a 10-to-1 scale model of the proposed designs and made just the same measurements, using tiny speakers, microphones, and one-inch "heads" of dummy audience members, all scaled down in proportion to the model. Even the wavelengths of the sound in the model measurements were scaled down.

This work led to numerous adjustments in the original designs, including changes in the height of the ceiling near the stage in the concert hall, giving some of the balcony fronts a rakish, forward slant and adding a special sound-diffusing material to the pyramidal ceiling.

Follow the Moving Sensor

Computer animators often use motion capture to transfer an actor's movements to an animated character. One common type of motion-capture system uses sensors and electromagnetic fields for tracking.

Polhemus

ACTING IT OUT

Actors wear sensors that are on special suits or strapped to the body. Sometimes they wear special gloves for tracking individual finger movements. Because magnetic fields are involved, motion-capture experts try to keep the performance area free of conductive metals that could produce distortion.

Dotcomix

PUTTING IT ALL TOGETHER

An actor's movements are tracked by the system and converted into data that is fed into a program that applies it to an animated character, shown here superimposed on the image.

TRANSMITTER

A globelike transmitter contains three coils that create electromagnetic fields along three axes. The fields induce voltages in coils contained in the sensors, much as in a transformer with two coils wrapped around a core, one coil induces a voltage in the other. In effect, this system is like a transformer in which the core is made of air.

SENSOR

Each sensor is actually a receiver that contains coils. The induced voltages vary depending upon how the sensor coils are oriented in the magnetic fields created by the transmitter.

BODY PACK

The data from the sensors can be sent to the computer system via cables, or wirelessly using a body pack that takes the signals and transmits them to the PC over radio waves.

APPLYING THE DATA

The position and orientation data is then used with a program that applies it to an animated character, to move it in the same way the actor moved.

DETERMINING POSITION

The voltage data from the sensors arrive at the PC, where it is used to determine the position and orientation of each sensor at any given point in time.

JOYSTICK

A joystick or other device can be used to control the movements of one or two specific features, like the eyes or an eyebrow.

LIP SYNCING

Software also can take the speech uttered by an actor and convert it to mouth movements for the animated character.

Sources: Polhemus Inc.; Dotcomix

Mika Gröndahl

Motion Capture Gives Cartoon Characters a Human Touch

By David Kushner

The presidential candidate was weary. He had made the interview rounds. Fielded the tough questions. Appeared on "Larry King Live." It's tough for anyone to remain animated under such scrutiny, especially for a digital creation like Uncle Duke, the mordant Doonesbury character who, thanks to an innovative new motion-capture technology, made a precipitous leap from the drawing board to live Web and television appearances.

For the sake of a satirical election campaign, the cartoonist Garry Trudeau's creation underwent a contemporary transformation from his two-dimensional life. Duke was first rendered as a three-dimensional computer-animated character, then evolved into a more realistic-looking character with the help of an actor in a motion-capture suit.

Given that injection of virtual and actual life, Duke could interact as spontaneously as his organic opponents. He streamed into PC's via the Internet and even took calls on television talk shows.

Magic like this isn't simply a byproduct of the ordinary motion-capture technology used in movie and computer game production — the kind that collects data from a real person's movements and applies that information to the movements of animated characters. Instead, it's digital puppetry, said Brad deGraf, founder and chief creative officer of Dotcomix, the animation studio behind the Duke wizardry.

Mr. DeGraf first explored the technology in 1987 while working with the Jim Henson, who was the creator of the Muppets and died in 1990.

Mr. Henson had an idea, Mr. deGraf said, to take what he was doing with remote control puppetry and replace it with a digital virtual one.

Until that point, motion-capture technology had largely been the domain of military and moviemaking science. The military had been using the technology since the 1970's so that, for example, a pilot in a specially equipped motion-capture helmet could track the motion of his guns by simply turning his head.

Bill Panepinto, vice president and director for sales and marketing at Polhemus, a company based in Colchester, Vt., that developed these applications, said that it had eventually become clear that the technology could be used to track body movements. The applications then migrated to Hollywood, where motion capture was used to add lifelike nuances to animated characters in films like "Beauty and the Beast" and "Toy Story."

In the 1990's, Mr. deGraf worked with others to expand the motion-capture tools to an environment that was more fluid and spontaneous. The resulting software achieves a kind of Frankenstein magic and is called, appropriately enough, Alive. It is the brain that pulls together a variety of motion-capture information.

Making a character like Duke come alive begins by sketching the character on a computer. When the character is completed, Mr. deGraf said, it is something like a marionette without the strings.

Attaching the strings that bring the character to life involves putting an actor into a body suit studded with nearly a dozen electromagnetic sensors. As the actor moves, the sensors travel within electromagnetic fields produced by a transmitter that resembles a disco ball.

The system collects data about the position and orientation of the sensors and uses that information to determine the sensors' location at any given time.

That data is then sent through a PC and into a Silicon Graphics workstation running the Alive software. At the same time, another puppeteer mimics the actor's facial expressions by using a special joystick. All the while, a lip-sync program is automatically converting the actor's speech into corresponding mouth movements.

Alive takes all the incoming data and applies it to Duke, essentially giving him his strings. For a television appearance, an actor can then respond to questions from an interviewer so the television audience sees an animated Duke moving seamlessly.

Fooling the Brain

A person perceives the world as three-dimensional because each eye transmits a slightly different view of the world to the brain. To make 3-D films, IMAX uses technological sleight of hand to slightly skew the images seen by the moviegoer's left and right eyes.

Shooting the Film

THE CAMERA

The 250-pound 3-D film camera is really two cameras in one, with two apertures that are about the same distance apart as a pair of human eyes. The images are recorded on two rolls of extra-large 70-millimeter film.

VIEWFINDER

SEPARATE OPTICS FOR BOTH ROLLS OF FILM

FILM FRAME FOR IMAX PROJECTOR IN ACTUAL SIZE

Showing the Film

LIQUID CRYSTAL LENS METHOD

A mechanical shutter in the projector alternates the images between both rolls of film so only the right-eye or left-eye image is on the screen at any given time.

LIQUID CRYSTAL VIEWERS

Liquid crystal lenses in the viewer become opaque or transparent in sync with the projector's shutter. So the right eye sees only right-eye images, the left eye only left-eye images.

THE PROJECTOR

Like the camera, the IMAX 3-D projector is really two projectors, one for the left-eye image, one for the right-eye image. The film spools lie flat.

A SENSE OF DEPTH

The brain combines the left and right images as if they were one. The slight differences between the two images helps create the illusion of depth.

INFRARED TRANSMITTER

Transmitter sends sync signals to the liquid crystal viewers.

PROJECTION BOOTH

The projection booth measures 21 by 76 feet, large enough to hold the nearly 1,500-pound projector and its cooling system.

SPEAKERS

A digital surround sound system creates clear sound throughout the theater.

THE SCREEN

The typical IMAX screen measures 95 feet wide by 71 feet high.

POLARIZED METHOD

The projector shows both left and right images simultaneously, but they are projected through different polarizing filters, which allow only light waves that are parallel to one plane to pass through.

Source: IMAX

POLARIZING VIEWER

The viewer has a polarizing filter for each eye. One of them is a vertical polarizer, and the other is a horizontal polarizer. Corresponding filters are used in the projector for the left-eye picture and right-eye picture. So the filter in the viewer for the right eye lets through only the right-eye picture, and the same goes for the left eye.

Mika Gröndahl

Movies with a Third Dimension That Makes You Duck

By David Kushner

Digital artists love to bandy about the term 3-D. There are so-called 3-D computer games. There are graphics called 3-D in films, like those used to anthropomorphize inanimate objects. In reality, though, these are little more than two-dimensional images with depth and shadows. They don't jump off the screen.

To get true in-your-face 3-D, more technology—usually involving some form of audience participation—is required.

The IMAX Corporation, the Canadian company that has developed the technology and theaters behind the over-size-movie experience, has been showing some of its films in 3-D since 1985, when "We Are Born of Stars" was previewed in Japan. Since then, nearly 80 3-D IMAX theaters have opened around the world, from Chattanooga, Tennessee, to Bangkok.

Thus far, the films have tried not only to make viewers duck their heads but also to shed some tears. "T-Rex: Back to the Cretaceous," released in 1998, followed the adventures of the teenage daughter of a paleontologist as she is transported 65 million years back in time. "Across the Sea of Time," from 1995, took the audience on a poignant tour of old New York by creating sweeping versions of vintage stereoscopic photos from the turn of the century.

Stereoscopic images, twin pictures viewed through twin lenses, ushered in the three-dimensional era in 1838. The first 3-D films came about in the early 1920s when movies were shown in anaglyphic format, which incorporated two slightly different images of a scene, dyed red and green, on the same film. When viewed through special glasses with a red filter for one eye and a green filter for the other, the images would come together as one and appear to have depth. This rather low-tech process peaked in the early 1950s with B movies like "Bwana Devil" and "House of Wax."

A live-action IMAX 3-D film is created by shooting the scenes with a special camera that has two apertures. By positioning the apertures at approximately the same distance as the distance between an average person's eyes, an image can be simultaneously recorded from slightly different perspectives on two separate rolls of film.

The film is then played back on a double-lensed projector that casts the image on the giant aluminum IMAX screen. Special shutters, filters, and viewing glasses let the viewer see the left and right images separately; the brain combines them to perceive the movie as three-dimensional.

Although the projection and the viewing process are identical for an animated film and for a live-action one, the 3-D aspect of an animated film is created in the ones and zeros of program code. Computer-animated projects like "Toy Story" are created using sophisticated 3-D modeling software. The IMAX conversion process essentially goes back into this software and shifts the point of view slightly to create two images that are like those used in live-action 3-D. By viewing the images simultaneously through the special glasses, the 3-D magic comes to life.

For animated clips made originally by IMAX, computer artists draw the animation with a special stereoscopic device that lets them draw digital images with a freely moving wand. The movements are than fed into a computer and stored as 3-D images. Making an IMAX film from scratch is time consuming and expensive; it could cost twice as much as it would take to convert a film with computer animation into the IMAX format.

The conversion technology could even be applied to the computer-generated effects in recent action and science-fiction films. And Bradley Wechsler, cochief executive of IMAX, speculated that someday similar technology could be used to transform live-action film sequences. If a colorized "Casablanca" looks bizarre, just imagine the possibilities of Bogey in 3-D. Here's really looking at you.

The Light Fantastic

Laser light shows, which first became popular in the 1970's, use a combination of lenses and mirrors to play colored beams on a screen or ceiling.

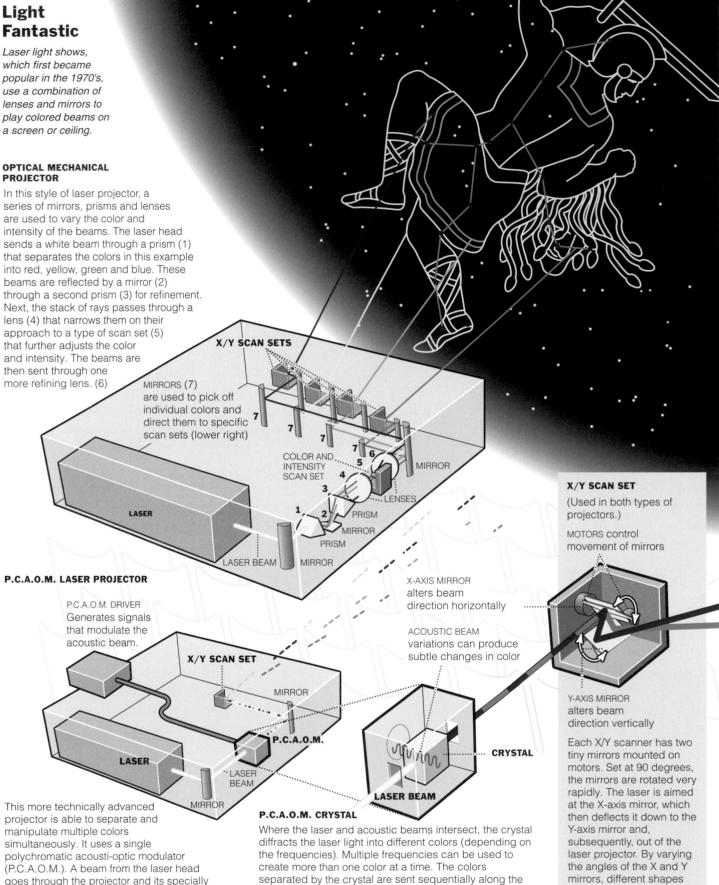

OPTICAL MECHANICAL PROJECTOR

In this style of laser projector, a series of mirrors, prisms and lenses are used to vary the color and intensity of the beams. The laser head sends a white beam through a prism (1) that separates the colors in this example into red, yellow, green and blue. These beams are reflected by a mirror (2) through a second prism (3) for refinement. Next, the stack of rays passes through a lens (4) that narrows them on their approach to a type of scan set (5) that further adjusts the color and intensity. The beams are then sent through one more refining lens. (6)

MIRRORS (7) are used to pick off individual colors and direct them to specific scan sets (lower right)

X/Y SCAN SETS

COLOR AND INTENSITY SCAN SET

MIRROR

LENSES

PRISM

MIRROR

PRISM

LASER

LASER BEAM

MIRROR

P.C.A.O.M. LASER PROJECTOR

P.C.A.O.M. DRIVER Generates signals that modulate the acoustic beam.

X/Y SCAN SET

MIRROR

P.C.A.O.M.

LASER

LASER BEAM

MIRROR

This more technically advanced projector is able to separate and manipulate multiple colors simultaneously. It uses a single polychromatic acousti-optic modulator (P.C.A.O.M.). A beam from the laser head goes through the projector and its specially vibrating internal crystal.

ACOUSTIC BEAM variations can produce subtle changes in color

CRYSTAL

LASER BEAM

P.C.A.O.M. CRYSTAL

Where the laser and acoustic beams intersect, the crystal diffracts the laser light into different colors (depending on the frequencies). Multiple frequencies can be used to create more than one color at a time. The colors separated by the crystal are sent sequentially along the same path. If, for example, a scan set is drawing in red, the P.C.A.O.M. will emit only red while it is required.

X/Y SCAN SET

(Used in both types of projectors.)

MOTORS control movement of mirrors

X-AXIS MIRROR alters beam direction horizontally

Y-AXIS MIRROR alters beam direction vertically

Each X/Y scanner has two tiny mirrors mounted on motors. Set at 90 degrees, the mirrors are rotated very rapidly. The laser is aimed at the X-axis mirror, which then deflects it down to the Y-axis mirror and, subsequently, out of the laser projector. By varying the angles of the X and Y mirrors, different shapes and images can be drawn.

Source: Laser Fantasy International

Science That Entertains: The Laser Puts on a Show

By David Kushner

As rockers lean back to gaze at a planetarium's dome, they see a glowing image pulsing in time with the drumbeat of the music. A figure of a hippie wearing neon red bell-bottoms and playing a quivering electric guitar disintegrates in a loud crash of cymbals and is replaced by a fluorescent wash of abstract designs that coalesce into luminous rising steps.

"Stairway to Heaven" by Led Zeppelin has finished to celebratory fist-pumping by the audience, ending this laser light show.

By most accounts, the laser show era was an outgrowth of the fascination of a California filmmaker, Ivan Dryer, with the laser experiments of a physicist in 1970. When Mr. Dryer documented the work, he said, he found himself "confronted by an eerie sci-fi environment." So Mr. Dryer, the unofficial father of laser light shows, began his own laser productions, including a stint creating laser effects for an Alice Cooper concert tour.

He started a company called Laser Images in 1971 and began presenting his own shows, set to recorded music — he dubbed them Laserium shows — in the planetarium at Griffith Observatory in Los Angeles. While Pink Floyd's "Dark Side of the Moon" played, for example, fans watched a display of laser lights choreographed to fit the moods and themes of the music.

Laser shows developed during the analog era, long before virtual reality, three-dimensional computer games and the Web. Now computers make it easier to create the images that form the patterns for the shows.

The International Laser Display Association said many companies now produced laser light shows at planetariums and science centers across the nation. The shows are also popular internationally, in places like Taiwan and Mexico.

Laser light shows are still the high-tech art of smoke and mirrors. Laser Fantasy of Redmond, Wash., one of the largest producers, says the process begins when computer animators and technicians produce a sequence of laser images that are synchronized with music. The laser images are produced by sending a laser beam through a sequence of lenses and mirrors.

For a live performance, a "laserist" will manipulate these programmed images by increasing their intensity, mixing colors or sending them through a fog machine's haze. During a guitar solo, wavy lines might tremble in time with the rhythm, or an outlined image of a sailing boat — reproduced from a series of computer-generated animation cells — could float overhead. "It's like watching a big spirograph that's being improvised like jazz," said Scott Huggins, director of sales for Laser Fantasy, referring to a popular art toy from the 1970's that is used to create complex spiral patterns.

The business faces increasing competition for audiences from everything from Imax theaters to blockbuster special effects in movies. "We're going to have to adapt or change," said Jack Dunn, who focuses on science centers for the International Laser Display Association. "Or planetariums will continue to be in trouble."

Mr. Dryer predicts that the laser shows' future will be in large-screen formats. The shows of the future will be displayed in Imax-style theaters that will combine traditional films with live laser effects, he said. Still, for many rock fans, nothing will replace the simple thrill of laser Zeppelin as it exists today.

The Latest Fashion Accessory

An electronic locating system developed by two companies is designed to enable visitors to amusement parks to keep track of family members or friends. Visitors can rent wrist-mounted transmitters that emit signals that are constantly monitored by antennas around the park. The location of each transmitter is determined by triangulation and displayed on electronic maps. Here is how the system works at a water park near Denver.

A TRANSMITTER ON THE WRIST

At the heart of the system is a watch-size waterproof transmitter worn by each member of a family or group. Each transmitter emits a signal with a unique identifying code.

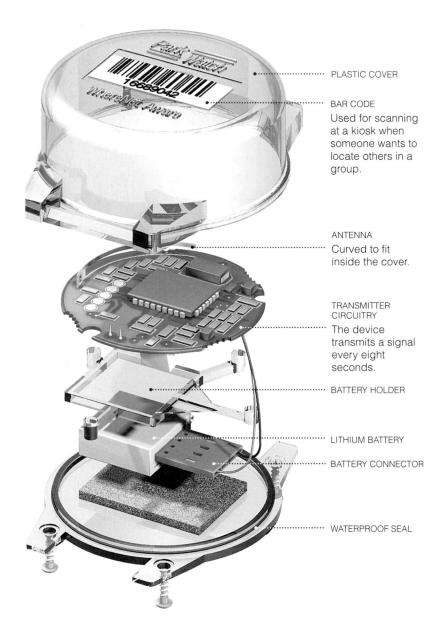

PLASTIC COVER

BAR CODE
Used for scanning at a kiosk when someone wants to locate others in a group.

ANTENNA
Curved to fit inside the cover.

TRANSMITTER CIRCUITRY
The device transmits a signal every eight seconds.

BATTERY HOLDER

LITHIUM BATTERY

BATTERY CONNECTOR

WATERPROOF SEAL

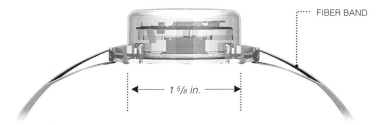

FIBER BAND

1 ⁵/₈ in.

PUTTING YOURSELF ON THE MAP

A user who wants to locate other family or group members goes to a kiosk that includes a scanner and an electronic map of the park. The scanner reads the bar code off the user's device, and then the locations of the other members of the party pop up on the screen.

ParkWatch Kiosk

ParkWatch User

THE PROCEDURE

1 Families are issued devices with linked I.D. numbers.

2 Devices send pulses to antennas around the park.

3 The computer uses triangulation to determine locations.

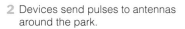

4 At a kiosk, a group member scans in the bar code from a wrist device.

5 The names and locations of all group members pop up on a map.

ON THE GO

Park employees carry wireless portable units that can also be used to track group members.

Sources: ParkWatch; WhereNet

Frank O'Connell

Personal Tracking Device: Something to Watch Over Me

By Mindy Sink

Electronic surveillance, family style, can be found at Hyland Hills Water World, a theme park north of downtown Denver.

On a typical summer day, thousands of children cavort among 65 acres of water slides, wave pools and rafting rides, and their parents try to keep up with them. An electronic locating system helps parents do just that.

Visitors can wear watch-size devices that transmit signals to antennas throughout the park. Family members can go to special kiosks in the park to have the whereabouts of others in their group displayed on a video map.

"It allows all of us a little bit more freedom and flexibility," said Chery Wick of Denver, who, like many parents, had never been comfortable allowing her children to roam freely in a theme park. But she rested contentedly by a pool at Water World, knowing that while her daughters, ages 9 and 12, were nowhere to be seen, she could find them because they were wearing the devices.

"They wouldn't be on their own right now without it," Ms. Wick said.

The system, developed by ParkWatch of Clarksburg, Md., and WhereNet of Santa Clara, Calif., is adapted from technology used to track parts in warehouses and factories.

It is also a precursor of systems that could track a person just about anywhere. Small tracking devices could be implanted under the skin or attached to a wristwatch and would use Global Positioning System chips and transmitters to determine and transmit the wearer's location.

But not everyone believes that the tracking devices are a good idea.

"There are a myriad of ethical and security issues," said Andrew Shen, a policy analyst for the Electronic Privacy Information Center in Washington.

"I'd worry about the people with malicious intent," Mr. Shen said, "and opening yourself up to abuses from people working at those companies where the tracking takes place."

Even a locator system in an amusement park, he added, "shouldn't be a substitute for keeping a watchful eye on a kid."

"Yes, you may be able to find your son or daughter," he added, "but that doesn't mean your kid is going to stay out of trouble. Kids can get hurt or upset if they are lost, and wearing a tracking device on their wrist doesn't help that much."

The developers of ParkWatch note that the use of the system is voluntary so security concerns are not a problem. People can choose to rent the devices for a daily fee.

"Anytime you have a large crowd, there is a possibility of separation," said Joanne Dobbs, communications manager for the park. "This can give parents peace of mind."

The high-tech aspect of the system has a certain appeal.

"I think it's pretty cool," said Gabrielle Villanueva, 8, as she raced toward a rafting ride at Water World. "It's keeping us safe even if we are separated."

The park has 14 antennas, many of them hidden, around the perimeter and six kiosks for calling up maps. Each tracking device weighs only a few ounces and is attached to the wrist, ankle or bathing suit with a fiber band that can be removed only by cutting it.

Each device has a serial number, and users can assign names or other labels.

To locate someone, anyone in the group can go to one of the electronic maps and scan a bar code on the wrist device. Within seconds, icons corresponding to the other members will appear on the screen, with their names or other labels.

Essentially, the device sends out a signal, with a unique code for each device, every eight seconds. The signal is picked up by the antennas, and then a computer uses triangulation (based on the slight time differences of the signal reaching the various antennas) to determine location within about 10 feet. The devices do not work underwater.

Dan Doles, president of WhereNet, compared ParkWatch with cell phones. "There are towers or antennas where you are within a coverage area," Mr. Doles said. "Except a cell phone has about 1 watt, and ParkWatch has only 2 to 3 milliwatts."

"I liken it to upside-down G.P.S.," Mr. O'Neill said. "In a G.P.S. environment, you know where it is from the satellite. Here it sends out a signal after it identifies itself."

Business
and
Industry

A Perfect Sense of Timing

The Global Positioning System, developed by the Defense Department, works by triangulation: calculating the position of an object based on its distance from several other objects. In this case, the reference objects are satellites. A G.P.S. receiver determines distance from the satellites by calculating the time it takes for signals to reach it.

THE SATELLITES
Powered by solar panels and containing highly accurate atomic clocks, the satellites transmit timing signals that are picked up by G.P.S. receivers.

COVERAGE
The system uses 24 satellites, which circle Earth every 12 hours at an altitude of about 12,000 miles.

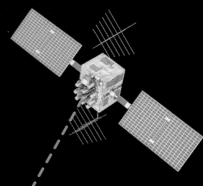

How triangulation works

COMPARING SIGNALS
A G.P.S. receiver generates a constantly changing code. Each satellite transmits its signal using the same codes, generated at the same time. The satellite signal is received later due to the distance traveled. The receiver determines the amount of delay by delaying its own code until it matches the satellite's.

1 With a signal from one satellite, the receiver could be at any point on a sphere (one that has the satellite at its center).

2 A signal from a second satellite narrows the possible location to a circle, where the two spheres intersect.

3 A signal from a third satellite narrows the location to two points, one of which can be rejected as nonsensical (out in space, for example).

4 A signal from a fourth satellite corrects for timing errors, increasing accuracy and effectively synchronizing a G.P.S. receiver's clock with the satellites'.

RECEIVER CODE

SATELLITE CODE

AMOUNT OF DELAY

SOLAR PANEL

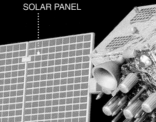

ANTENNA ARRAY

GROUND STATIONS
Because satellite orbits vary slightly over time, a network of ground stations monitors the orbits and sends updated position information to each satellite along with the timing signals.

Source: Lockheed Martin Corporation

Frank O'Connell

The Global Positioning System: Getting There With Help From Above

By Matt Lake

In the early 1970's, the Defense Department needed a navigational tool that troops on the move could use to pinpoint their location. The solution they developed required two dozen satellites, atomic clocks, microwave radio transmitters and some heavy-duty number-crunching hardware.

The military called it the Global Positioning System, or G.P.S., and like the Internet, it was a cold war development that is now used by millions of civilians for fun and profit. Any time you take a plane trip, receive goods from a commercial shipping agency or drive a car with a NeverLost or OnStar navigation system, you are benefiting from G.P.S. technology.

With G.P.S., a receiver on the ground or in the air can calculate its position using time signals from the satellites. The calculation itself is based on a kind of triangulation — a high-school math technique used to locate an object based on its distance from three points. So signals from three satellites are necessary, although in practice a signal from a fourth satellite is used to improve the accuracy of the other three signals.

To calculate these distances, the system uses another basic high-school math equation: distance is equal to the speed of travel multiplied by the time.

In addition to the time, a signal from a G.P.S. satellite also includes information about the satellite's exact location, which is known, tracked and kept accurate by ground control stations. The time signal is also very accurate, because each satellite contains several atomic clocks. These rely on the natural, and very regular, oscillation frequencies of atoms to keep time.

The end result is that a G.P.S. receiver — perhaps an in-dash navigation system in a car — can produce highly accurate coordinates of latitude, longitude and altitude.

The final step, placing your location on a map and providing directions, falls to software developers in the commercial sector. Companies like Magellan of Santa Clara, Calif., and TravRoute of Princeton, N.J., provide consumer products that combine a G.P.S. receiver with map programs that can provide turn-by-turn directions for drivers. From longitude and latitude readings that update every second, these programs can determine the speed and direction, information that enables them to superimpose a "you are here" arrow on the display of a road map.

This navigation technology also has implications for the trucking industry, which has begun to automate its fleet management systems with customized G.P.S. equipment. These heavy-duty systems not only receive G.P.S. satellite signals and calculate location, they also transmit location information to a central computer that handles route planning for multiple deliveries. This enables new delivery and pickup points to be added to a driver's route without requiring any data entry by the driver.

The G.P.S. does have some blind spots, however. Alain Kornhauser, director of the interdepartmental transportation research program at Princeton University and founder of TravRoute, identifies trees, tunnels, power lines and tall buildings as stumbling blocks for satellite signals.

"Wall Street is a problem area," Dr. Kornhauser said. "It is really narrow with tall buildings."

Another problem is the orbits of the satellites, which must pass within range of the system's five ground control stations, so not all parts of the earth have the minimum four satellites within range at all times.

A greater source of error is atmospheric conditions, which can ever-so-slightly slow down the radio signals from the satellites, introducing errors into the system.

To help improve accuracy, an enhancement, called differential G.P.S., uses stationary ground-based receivers to monitor the signals from the constellation of G.P.S. satellites. Because the precise locations of differential G.P.S. stations are known, they can be used to correct errors in the system. These stations then transmit corrective formulas that special differential G.P.S. receivers can receive and incorporate into their calculations. Using this trick, differential G.P.S. can provide accuracy down to a meter or so, which is all the more impressive when you consider that the G.P.S. satellites are some 12,000 miles above the earth.

A Power Plant in the Sky

A wind turbine converts energy from moving air into electrical energy. Although modern turbines are designed to be highly efficient, some of the energy is lost to heat and sound. Here's a look inside one of the largest wind turbines made, one that produces about 1.5 megawatts of electricity.

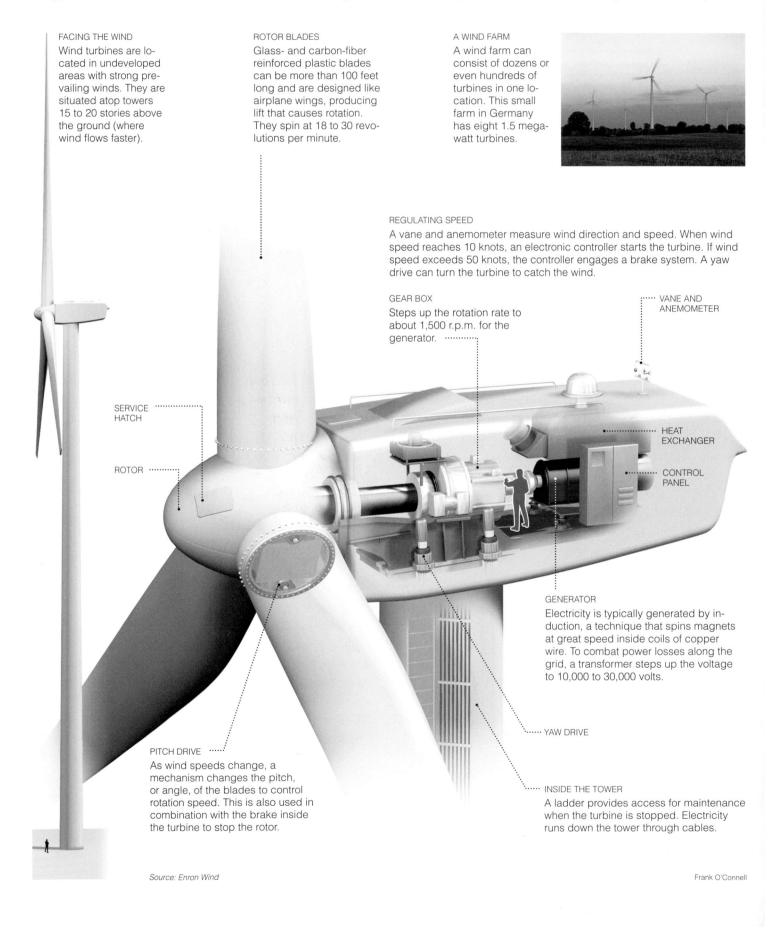

FACING THE WIND
Wind turbines are located in undeveloped areas with strong prevailing winds. They are situated atop towers 15 to 20 stories above the ground (where wind flows faster).

ROTOR BLADES
Glass- and carbon-fiber reinforced plastic blades can be more than 100 feet long and are designed like airplane wings, producing lift that causes rotation. They spin at 18 to 30 revolutions per minute.

A WIND FARM
A wind farm can consist of dozens or even hundreds of turbines in one location. This small farm in Germany has eight 1.5 megawatt turbines.

REGULATING SPEED
A vane and anemometer measure wind direction and speed. When wind speed reaches 10 knots, an electronic controller starts the turbine. If wind speed exceeds 50 knots, the controller engages a brake system. A yaw drive can turn the turbine to catch the wind.

GEAR BOX
Steps up the rotation rate to about 1,500 r.p.m. for the generator.

VANE AND ANEMOMETER

SERVICE HATCH

HEAT EXCHANGER

ROTOR

CONTROL PANEL

GENERATOR
Electricity is typically generated by induction, a technique that spins magnets at great speed inside coils of copper wire. To combat power losses along the grid, a transformer steps up the voltage to 10,000 to 30,000 volts.

YAW DRIVE

PITCH DRIVE
As wind speeds change, a mechanism changes the pitch, or angle, of the blades to control rotation speed. This is also used in combination with the brake inside the turbine to stop the rotor.

INSIDE THE TOWER
A ladder provides access for maintenance when the turbine is stopped. Electricity runs down the tower through cables.

Source: Enron Wind

Frank O'Connell

On Energy Farms, Giant Pinwheels Harvest the Wind

By Matt Lake

The windmill is a centuries-old technology originally developed to pump water and grind grain, but it could play an important role in solving two serious problems facing the United States: generating an adequate supply of electricity and reducing the emission of greenhouse gases.

Wind turbines have been used for generating electricity for more than a century, beginning with experiments by a Danish meteorologist, Poul la Cour, in the 1890's and his founding of the Society of Wind Electricians in 1905.

Despite refinements in the 20th century, development did not begin in earnest until the oil crisis in the 1970's made the electricity supply expensive and less reliable. The 1970's and 1980's saw a proliferation of turbine designs, ranging from two- and three-propeller horizontal generator designs to a vertical model that resembled an eggbeater. But this country's nascent wind generation industry got a bad rap in the 1980's when investments in wind generators failed because of cheap electricity from fossil and nuclear fuels and the inefficient design of some turbines.

Wind turbines have now become much more efficient, thanks in large part to the Danish windmill industry. Now dozens of manufacturers are cranking out turbine models, from small designs, aimed at providing power for a single house, to huge machines with 100-foot blades that can supply between two and three million kilowatt-hours in a year, enough to power at least 500 households.

California, which has the most developed wind-power industry of any state, has more than 15,000 turbines creating enough energy for about a million people. And in the late 1990's, Enron Wind, a wind-turbine developer based in California, installed some of the largest wind-generation "farms" in the world in the American Midwest: two plants, each producing more than a hundred megawatts, in Lake Benton, Minn., and a 193-megawatt plant in Storm Lake, Iowa.

The impetus for generating clean power is being spurred by efforts to reduce greenhouse gases. Currently, two-thirds of the electricity in the United States is generated by burning fossil fuels like coal, gas and oil. Such combustion pumps more than 2 million metric tons of carbon dioxide into the atmosphere, according to the Department of Energy and the Environmental Protection Agency.

The more power produced by wind, the less need be produced by burning fossil fuels. But wind turbines remain a relatively untapped resource, contributing less than 1 percent of the national supply.

A disadvantage of wind power is its unreliability. Although wind farms are situated to take advantage of strong prevailing winds, variations in wind speed cause unpredictable fluctuations in the capacity of a wind turbine. Even the most efficient designs fail to capture 40 percent of the energy in the wind. But since it's "free" energy, failing to tap it all is not such a big problem.

Despite the advantages of wind generation, it has a large number of earnest opponents. Some people just do not want to have wind turbines nearby. The turbines are huge structures that are visible from great distances because they need unobstructed access to air currents. So many zoning boards and residents near a proposed project say they are eyesores that contribute to visual blight.

The large structures also require deep concrete foundations, and that heavy construction has its own impact on rural settings. Such excavation leads to local air pollution and the risk of erosion at the sites and their access roads. And wind farms gobble up territory, taking up considerably more land than other electrical generators to produce a comparable amount of electricity.

Many older models drew opposition because they were noisy, but newer designs are quieter. The wind generation industry says modern turbines are no louder than refrigerators when heard from the distance that most turbines are placed from populated areas.

There is also controversy over the impact of wind turbines on the environment, especially bird life, because raptors like golden eagles are drawn to prey sheltering near the turbines and can be killed by the spinning blades. The Sierra Club, the United States Fish and Wildlife Service, and the Audubon Society all recognize bird mortality as a significant problem with wind generators.

Gas In, Electrons Out

A fuel cell is an efficient and clean (if costly) way of making electricity. It works somewhat like a battery, but it has a constant source of fuel — hydrogen and oxygen, which combine in the presence of electrolytes. Most fuel cells use oxygen from the air and hydrogen from a fuel like natural gas. Many different kinds of electrolytes can be used; the system pictured here, a 200-kilowatt model, uses phosphoric acid.

FUEL CELL SYSTEM

The actual fuel cells form only a part of a complete power system. Among other components, the truck-size unit contains a reformer to produce hydrogen from the natural gas fuel, a shift converter that removes carbon monoxide from the hydrogen gas mixture, a cooling system, a condenser for hot gases and electrical equipment.

FUEL CELL STACK

Individual cells produce only a small voltage; connecting them in a stack increases the voltage. Current is determined by the area of each cell, and current times voltage equals total wattage, or power. This system has 272 cells and produces 200,000 watts.

MAINTAINING PRESSURE

Manifolds on outside of stack keep the gases under pressure so they flow through the cells.

INSIDE A CELL

Hydrogen gas enters through channels and permeates into porous anode; air does the same through the cathode. Both electrodes are coated with platinum, which acts as a catalyst. The electrolyte, phosphoric acid, is sandwiched between the electrodes. There is one cooling plate for every eight cells.

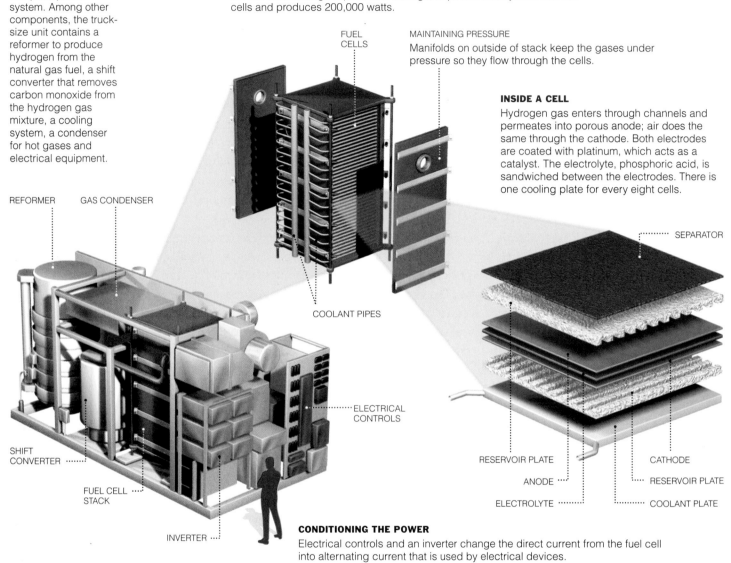

FUEL CELLS

SEPARATOR

COOLANT PIPES

REFORMER   GAS CONDENSER

ELECTRICAL CONTROLS

SHIFT CONVERTER

FUEL CELL STACK

INVERTER

RESERVOIR PLATE

ANODE

ELECTROLYTE

CATHODE

RESERVOIR PLATE

COOLANT PLATE

CONDITIONING THE POWER

Electrical controls and an inverter change the direct current from the fuel cell into alternating current that is used by electrical devices.

An Electrochemical Reaction

In the cells, the electrons are stripped off hydrogen atoms in the platinum catalyst. The electrons leave the cell, creating current.

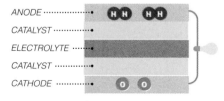

ANODE

CATALYST

ELECTROLYTE

CATALYST

CATHODE

1 Hydrogen travels through channels on a reservoir plate and permeates the anode. Oxygen (in air) permeates the cathode.

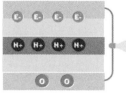

2 Hydrogen loses electrons, aided by a catalyst. Hydrogen ions (protons) pass into the electrolyte.

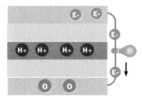

3 Electrons cannot enter the electrolyte; they travel through an external circuit, performing work.

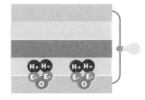

4 Electrons travel back to cathode, where they combine with hydrogen ions and oxygen to make water.

Source: International Fuel Cells

Frank O'Connell

Fuel Cells: Clean, Reliable (and Pricey) Electricity

By Catherine Greenman

If New York City had a power blackout, the Central Park police station would continue to glow like a beacon inside a dark forest. That's because the station gets all its electricity from a fuel cell right outside the building, not from utility lines. The fuel cell, which is about the size of a small delivery truck, produces a reassuring low hum in addition to 200,000 watts of electrical power.

Fuel cells are one of the cleanest and most efficient technologies for producing electricity. They turn oxygen and hydrogen into electricity in the presence of electrolytes, charged particles in solution. Fuel cells emit negligible amounts of pollutants. They work somewhat like batteries, in that both use electrolytes and electrodes to conduct current, but the cells use a constant source of fuel (usually natural gas for the hydrogen and air for the oxygen) so they don't lose their charge as batteries do.

Fuel cells are often seen as a reliable source of backup power. "We're really high on the technology," said Michael Saltzman, a spokesman for the New York Power Authority. "The end user has the generator capacity right at their doorstep, so there's no relying on transmission and distribution lines. And by virtue of that, you're more certain that you have reliable power."

Fuel cells were invented more than 150 years ago, although the first modern ones, producing useful amounts of power, were not developed until after World War II. NASA financed fuel cell projects in the 1960's and 70's in an effort to develop compact generators for its spacecraft. NASA's cells use tanks of liquefied hydrogen and liquefied oxygen as fuel (it was a fuel cell's oxygen tank that exploded during the Apollo 13 flight, crippling the mission).

Most of the commercially available fuel cells today use phosphoric acid as the electrolyte. Newer types use other materials. One of the most promising is called a proton-exchange membrane, or PEM, cell, and uses a polymer electrolyte membrane instead of an electrolyte in solution. PEM cells, which are being developed by International Fuel Cells and other companies, can be lighter than other kinds of fuel cells so they are being studied for use in cars and buses. They also operate at lower temperatures.

As well regarded as fuel cells are, very few companies and even fewer individual homeowners actually use them because of their high cost, three to five times higher per kilowatt than a diesel generator of similar capacity.

"For homes that average a kilowatt of power for a whole year, it's very difficult to save enough money by undercutting the cost of the utilities," said Dr. Jack Brouwer, the associate director of the National Fuel Cell Research Center at the University of California at Irvine. "It would be better to go into the stock market than to pay for the capital of this thing sitting in your basement and expect to save money."

Dr. Brouwer said the success of the fuel cell industry would depend on its ability to gain customers and manufacturing experience through niche markets that can pay more for reliable power. Credit card companies, for example, may be willing to purchase fuel cells, Dr. Brouwer said, because they can lose millions of dollars an hour if the power supply is interrupted. (First National Bank of Omaha uses four units from International Fuel Cells so it can have an independent power supply if the grid power goes out.)

Other potential markets include houses in remote areas beyond the reach of power lines. Plug Power, based in Latham, N.Y., is developing PEM fuel cells with maximum outputs of 150 kilowatts that are aimed at residential and light-commercial uses.

Dr. Tom Kreutz, a research scientist at the Center for Energy and Environmental Studies at Princeton University, said another potential niche market was in lightweight fuel cells for laptops and cell phones because some people would be willing to pay a premium for longer battery life. "The thing with fuel cells is they're very expensive so you've got to find some value proposition to sell them," Dr. Kreutz said. "Phosphoric acid and PEM marketers are looking for places where people want to pay for reliability and longer battery life."

Flip That Switch, and a Generator Works a Little Harder

The complex network that moves electricity from where it is made to where it is used is a model of cooperation. Transformers are used throughout the grid to increase voltage for transmission over long distances or reduce it for local distribution to homes and businesses.

GENERATING PLANTS

The most common types of generators use coal, natural gas, oil, falling water or nuclear energy to produce electricity. But garbage and other waste can also be burned to produce electrical energy, and generators can harness wind or solar power. Total generator output has to rise and fall to meet the total demands on the electric system, and a combination of types of generating units is used to match the demand. Economic considerations, environmental constraints and availability determine which units supply power at a given time. Different combinations of these power sources are most economical under different conditions, like the time of year. In the fall, for example, reservoirs tend to be depleted, so hydroelectric power is less readily available. Generators usually produce electricity at 7 kilovolts (a kilovolt is a thousand volts) to 25 kilovolts; transformers step up the voltage, to 60 kilovolts to 500 kilovolts, before sending the power into the grid.

TRANSMISSION SUBSTATION

Transmission substations act as way stations that link various lines for increased reliability. They can also step up the voltage again, depending on how far the electricity must travel.

TRANSMISSION LINES

High-voltage lines are more efficient for transmitting power. Raising the voltage lowers the current, and since wires waste power in proportion to the square of the current, that results in less power loss. Even so, substantial amounts of power are lost. (Those orange balls occasionally seen on high-tension lines, by the way, are to warn off aircraft.)

DISTRIBUTION SUBSTATION

At distribution substations, the electricity is stepped back down for customers, usually to between 12 kilovolts and 35 kilovolts, depending on how the individual system was designed.

DISTRIBUTION LINES

When the electricity is sent from a distribution line to a house, field transformers step down the power to 120-240 volts (the voltage that comes out of the plug in your wall). In commercial areas, the final voltages may be higher.

7-25 kilovolts

60-500 kilovolts

Typically 200 kilovolts

12-35 kilovolts

120-240 volts

TRANSFORMER

COILS

CORE

INSIDE A TRANSFORMER

A final step-down transformer, often located on a utility pole or in a vault under a city street, decreases the voltage to the 120-240 volts used in most homes. A transformer consists of two wires wound around a metal core. Power enters the first, or primary, coil and induces a magnetic field in the core, like an electromagnet. As the current alternates, the magnetic field reverses, creating an electric field that induces a current in the secondary coil. The ratio of the number of turns in the two coils determines the voltage change. If the secondary coil has half the turns of the primary, the voltage is stepped down by half; if it has twice the turns, the voltage is doubled.

MAJOR ELECTRICAL INTERCONNECTIONS OF NORTH AMERICA

CANADA

Western Interconnection

Eastern Interconnection

UNITED STATES

Ercot*

*Electric Reliability Council of Texas

Sources: Edison Electric Institute, California Independent System Operator, Southeastern Transformer Company

Mika Gröndahl

Getting Power to the People: The North American Electricity Grid

By Sally McGrane

You go home, unlock the front door and flip a switch, and your hallway lights up.

It may seem like a simple thing, turning on that 60-watt bulb. But that bulb is actually part of a web of generators, power lines and other people's bulbs that are interconnected in a complex electrical network known as the North American power grid. It stretches from one end of the United States to the other and includes parts of Canada and Mexico, with more than 700,000 miles of high-tension lines.

Within the grid are three interconnected divisions; within each one, the amount of electricity used must equal the amount of electricity produced at every given instant. Because the alternating current in your electrical lines travels at virtually the speed of light and cannot be stored, every time you turn on a reading lamp, a generator somewhere in the grid has to provide a tiny bit more power to the entire system.

The electricity travels long distances over circuitous routes, following paths of least resistance. It's like the flow of traffic into a city: if one freeway is jammed, cars will use alternate routes.

Take power shipped from Portland, Ore., to Los Angeles, for example: some 30 percent of that shipment will probably flow through Utah en route. Meanwhile in the East, if a Canadian hydropower plant is fueling someone's midnight repast in New York, the electricity could flow as far west as Ohio and as far south as Virginia — nearly instantly — before reaching the house where it is needed.

Power has to be routed, just like passenger or freight trains, and throughout the grid there are switches and distribution equipment to do this. But for efficiency, the electricity also must undergo several changes in voltage.

Relatively low-voltage power produced at the generating plant is stepped up by transformers to extremely high voltages — in the hundreds of thousands of volts — before it is sent through transmission lines. This reduces the power losses that occur when electricity is sent long distances.

At the other end, substation transformers step the voltage back down for local distribution. Smaller neighborhood transformers, often located at utility poles, reduce the voltage even further, usually to 240 volts, the typical house voltage.

If lightning knocks out one line in the grid, or a generator stops working unexpectedly, the rest of the grid is affected. Other lines handle more power to make up for the downed line, or other generators work harder. Every part of the system is sensitive to changes in any other part of the system.

While the system uses alternating current, or AC, the three divisions are connected by direct current, or DC, lines, which are easier to control. So in an emergency, power can be transferred from one connection to another, but power failures will not spread.

The power grid has evolved over the last hundred years or more. Gene Gorzelnik, director of communications at the North American Electric Reliability Council, said that 138 "control areas" monitored the modern grid, using computer systems to predict energy flow and anticipate reactions to power failures.

John Casazza, author of "The Development of Electric Power Transmission," remembers a simpler time. Mr. Casazza recalls working with an early analog computer, built in 1927, that used electric components to reproduce an electric system in miniature. Such early devices, each big enough to fill a large living room, were used until the 60's to simulate the impact of various events on the grid. In the 60's, electric companies switched to digital computers.

The first commercial power plant opened in San Francisco in 1879. That was followed in 1882 by Thomas Edison's better known Pearl Street station in New York, which delivered electric power using DC lines. In 1893, AC lines were displayed at the Chicago World's Fair, and in 1896, AC lines delivered energy harnessed at Niagara Falls some 20 miles to Buffalo, setting today's AC-line standard.

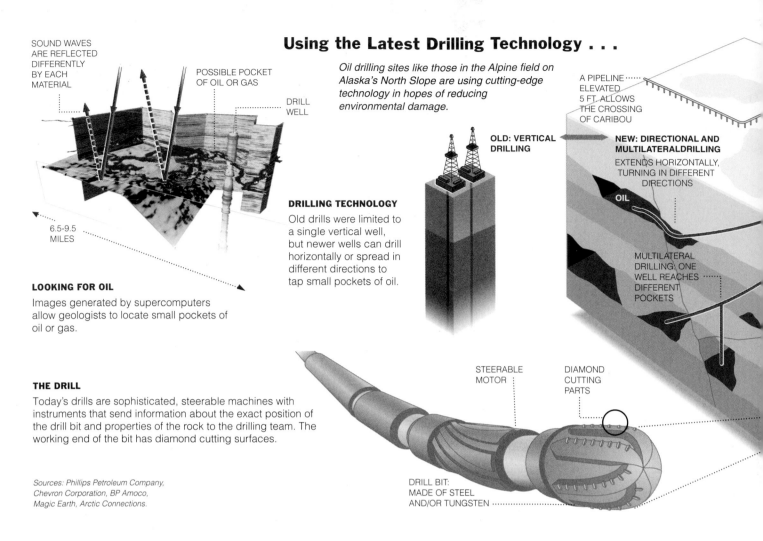

Using the Latest Drilling Technology . . .

SOUND WAVES ARE REFLECTED DIFFERENTLY BY EACH MATERIAL

POSSIBLE POCKET OF OIL OR GAS

DRILL WELL

Oil drilling sites like those in the Alpine field on Alaska's North Slope are using cutting-edge technology in hopes of reducing environmental damage.

A PIPELINE ELEVATED 5 FT. ALLOWS THE CROSSING OF CARIBOU

OLD: VERTICAL DRILLING

NEW: DIRECTIONAL AND MULTILATERALDRILLING
EXTENDS HORIZONTALLY, TURNING IN DIFFERENT DIRECTIONS

OIL

6.5-9.5 MILES

DRILLING TECHNOLOGY
Old drills were limited to a single vertical well, but newer wells can drill horizontally or spread in different directions to tap small pockets of oil.

MULTILATERAL DRILLING: ONE WELL REACHES DIFFERENT POCKETS

LOOKING FOR OIL
Images generated by supercomputers allow geologists to locate small pockets of oil or gas.

STEERABLE MOTOR

DIAMOND CUTTING PARTS

THE DRILL
Today's drills are sophisticated, steerable machines with instruments that send information about the exact position of the drill bit and properties of the rock to the drilling team. The working end of the bit has diamond cutting surfaces.

Sources: Phillips Petroleum Company, Chevron Corporation, BP Amoco, Magic Earth, Arctic Connections.

DRILL BIT: MADE OF STEEL AND/OR TUNGSTEN

Hunting for Oil: New Precision, Less Pollution

By Andrew C. Revkin

Each year, in the midwinter gloom of Alaska's Arctic, a rush begins to probe the North Slope for undiscovered pockets of oil. The ritual always starts like a horse race, when state officials declare the region open for "tundra travel." It means the normally fragile peat bogs, shallow ponds and lichen-tufted ridges are frozen as solid as a tabletop, and the seismic-mapping, road-building and well-drilling teams can fan out across the windblown vastness.

Although the rhythm of the work has been the same for years, the effort now bears little resemblance to the early days of the Alaskan oil rush, when the first finds at Prudhoe Bay in 1968 led the Trans-Alaska Pipeline and opened the country's last big terrestrial oil frontier.

Back then, thousands of tons of gravel were scooped from riverbeds and spread far and wide on the peat to make roads and drilling pads. Exploration wells sprouted across the treeless plain. Waste pits overflowed with drilling mud, contaminated water, spilled oil and discarded chemicals.

But since the early 1990's, the quest for Arctic oil has quietly, and almost completely, changed.

Oil drilling is still far from a green industry, but advances in computing and exploration methods, new techniques that allow dispersed underground targets to be reached from a single drilling site and different waste disposal practices have greatly reduced the environmental damage.

Supercomputer simulations of the deep earth are raising success rates, cutting the need to bore extra wells. New sensor-studded drills can steer through rock, sand and salt layers in three dimensions, following thin layers containing oil.

Wells are now constructed in tight ranks on small gravel pads, branching out underground instead of peppering the surface over a broad area.

Waste, garbage and cuttings of rock are ground into a slurry and pumped into sediment layers 2,000 feet beneath

. . . To Reduce Environmental Damage

ALPINE FIELD, ALASKA

DISPOSING OF OIL FIELD WASTES

DRILL

PERMAFROST

MUD LAYER

OIL LAYER

BP Amoco

MUD/ROCKPUMPING, CRUSHING AND INJECTION

DISPOSING OF WASTES
Mud and debris from drilling used to be placed in big reserve pits. Today, rock cuttings are crushed, mixed with the mud, and sent deep into the earth where they originated. This minimizes the size of well pads.

GETTING THERE
Roads to drilling sites used to be constructed of gravel, permanently scarring the landscape. Now, to minimize the impact on the environment, temporary ice roads are used in the winter, leaving few traces after they thaw.

LESS OF THE TUNDRA
The new drilling technology allows for smaller surface production pads and larger areas explored in the earth.

13 ACRES — 1999 — 8 MILES
11 ACRES — 1985 — 5 MILES
24 ACRES — 1980 — 3 MILES
5 ACRES — 1970 — 2 M
DRILLSITE SIZE
SUBSURFACE DRILLABLE AREA

Juan Velasco

the permafrost. Roads once built of gravel are now often built of ice and melt away in the spring thaw, leaving few traces on the spongy ground.

Critics and some environmental groups, while acknowledging improvements, say the oil quest always leaves scars, the innovations come with new risks, and — most important — once oil is found a wild landscape is permanently transformed.

The three-dimensional seismic technology that essentially performs ultrasound on the earth clearly is a mixed blessing, both industry and environmental critics say.

Although seismic crews now conduct surveys with vibrating trucks instead of detonated dynamite, their convoys, despite wide, tanklike treads that are designed to spread the weight of 10-ton vehicles, still can leave persistent scars even when traveling on frozen ground.

These high-tech wagon trains do not travel on ice roads, but instead crisscross the open tundra. They cover vastly greater acreage than the older seismic method so they can provide the most precise data possible for the three-dimensional simulators back in Anchorage or Houston. But it is also

much more likely that they will encounter a herd of musk oxen or a polar bear den.

In the meantime, there are significant environmental and economic advantages to the new seismic methods, which provide a vastly improved understanding of subterranean conditions. This is already greatly reducing the rate of "dry holes" — wells that are drilled but come up empty. The fewer wells drilled, the smaller the impact on the landscape.

The technology of drilling has improved at almost the same pace. The brute-force mechanics of drilling still involve pumping high-pressure mud down the hole to spin the grinding faces of the drill bit. But now the bit can be steered precisely in three dimensions so that a well can plunge down through the permafrost and into rock, and then head horizontally for four or five miles, threading its way through oil-bearing zones along the way.

New drills are studded with sensors that provide a steady stream of data on location, rock type, the presence of fluid, temperature and other conditions. That information is sent back to the surface coded in vibrations traveling through the stream of mud powering the drill.

Serving Up Data

Fast-food restaurants use computer networks to track ordering, cooking, serving and inventory. Some restaurants, like the Burger King depicted here, send data from their network to company headquarters daily or more frequently.

HEADQUARTERS IN MIAMI
Company-owned Burger Kings are linked to headquarters via a dial-up modem. Each night, all data on sales, inventory and time spent per order are uploaded to the central office.

DIVVYING UP THE ORDER

Orders for each food item are sent to the appropriate preparation spot in the restaurant. If a customer changes an order after it's been transmitted, the original order flashes to alert the workers filling the order.

AT THE COUNTER

A printer on the counter prints the order for the "expedite person," who assembles the non-burger part of the order.

AT THE GRILL

If a customer orders a cheeseburger with extra pickles and a small order of fries, the cheeseburger request appears on a monitor above the person cooking the burgers, along with the number of the order and the time it was placed. Orders for french fries are routed to the right station near the counter.

SEATING AREA

ENTRANCE

MANAGER'S OFFICE

NT SERVER

DRINKS

COMPUTER MONITOR

COUNTER COMPUTERS

FRIES

SINKS

BURGER PREPA- RATION

PRINTER

DATA CONNECTION

RESTROOMS

BROILER

14-INCH L.C.D. SCREEN

STORAGE

EXPEDITE PERSON

COLD STORAGE

PICKUP WINDOW

PAY WINDOW

DRIVE-THROUGH ORDERS

Drive-through orders take a similar route. Since the outside speakers are often unintelligible, a customer's order also appears on a video screen for accuracy.

COLOR CODING

Drive-through orders appear on the same video monitor as purchases in the store, but drive-through orders are in bright green type and the others are in blue. The food-preparation people need to know where the orders are coming from because they fill the orders placed from cars first. (The company doesn't want the line outside to get too long because that would discourage others from joining it.) By getting drive-through customers to pay at one window and pick up their orders at the next, the line keeps moving and people have the illusion that they are making faster progress.

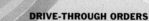

A program called Composer keeps track of basic restaurant data like food consumption and preparation time.

Source: Burger King; Compris Technology

Mika Gröndahl

The Fast-Food Industry's Secret Ingredient: Computers

By Eric A. Taub

Ah, the joys of youth. What baby boomer can't remember the drive with a date to the local Big Boy, A & W or Dairy Queen, where a young woman cheerfully took orders from people in parked cars, then delivered shakes and cheeseburgers she might have even helped to prepare?

Forget it. In the fast-food industry today, there's no room for individual expression, carefree service or pitching in to help when the going gets tough. The next time you chow down on one of the 1,024 permutations of a Burger King Whopper (the customer can make up to 10 choices per burger, like whether to have cheese or pickles), you might reflect on the fact that burger production has become an exact science in which everything is regimented, every distance calculated and every dollop of ketchup monitored and tracked.

"Each of our company-owned restaurants is a sophisticated network," said Patrick Forney, who oversees the sales-tracking systems at the Burger King Corporation's headquarters in Miami. Special software is used to monitor every aspect of the ordering and serving process in real time. That information is garnered from every counter sales terminal and then sent via a local network to a Windows NT server in the restaurant manager's office. Headquarters gets each store's data at least once a day.

Every major fast-food chain, including McDonald's, Hardee's, Kentucky Fried Chicken, Taco Bell and Wendy's, uses a version of the same software package in at least some of their stores, according to the company that makes the software, Compris Technologies.

Burger King's process begins the moment you're ready to buy. The counter person takes your order and types it into the terminal, including such special requests as extra pickles, no mustard or three slices of tomato. Each item that makes up that order is then automatically deleted from inventory.

A list of the items ordered is sent to the appropriate preparation station. Burger orders appear on a video screen near the grill, while the "expedite person" gets a printout showing the order for fries, desserts and shakes.

Pricing information is constantly updated in each terminal. If the chain is running a special, like half-price children's meals, then Burger King headquarters sends a command directly to the computers at every one of its company-owned stores to alter the menu for the period the special is being offered. Each terminal automatically reverts to the old price when the special offer expires.

To accomplish all that, the manager of each Burger King uses an off-the-shelf Windows NT server to transmit and receive data to and from the terminals at the sales counter and to store data. It runs a modified version of the restaurant software, called Composer, to track food consumption, the time needed to fill an order, the time needed to give change, how long it takes for an order to be entered and standard bookkeeping data on employees, like payroll and overtime information.

"With Composer, you can figure out how much of an item is going to be ordered during every upcoming 15-minute period, even distinguishing between days of the week, by averaging sales over the past nine weeks," said Paul Eurek, president of Compris, in Kennesaw, Ga.

The software monitors the use of condiments like ketchup down to units of 0.12 ounces and calculates when it is time to order more food. For example, the number of tomatoes used in a day is calculated by dividing the number of slices served by the typical number of slices per tomato; the total number of tomato slices served is automatically kept up to date each time an item is entered into a computer at the counter

Each evening, that day's sales data are uploaded from the store's server to the company's headquarters. The main office can also tap into a local company-owned store's computer at any time to see what's going on. Headquarters can even take over a counter terminal if something's going wrong.

All that helps fiercely competitive fast-food franchises make sure that food costs, as well as the food itself, are cut to the bone.

Life Down Under

One way to install high-capacity fiber optic cable in cities is to run it through existing sewer pipes. This gets the cable close to most buildings and avoids having to tear up local streets. Since sewer pipes are narrow, a remote-controlled robot does the work.

MANHOLE ··········

SEWER PIPE · · · · · · ROBOT

ON THE STREET
Workers lower the robot into the manhole, insert it into the sewer pipe and monitor and control its operation. All the robot modules include video cameras so that operators can see what is going on.

DOWN UNDER
The work takes place at night, when there is less flow in the sewers. The pipe is washed first with high-pressure water, then the robot makes at least three passes through it. On a good night, the machine can install conduit in about 300 feet of pipe.

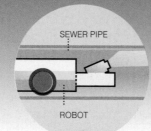

SEWER PIPE

ROBOT

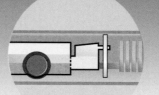

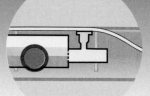

1 MAPPING
The robot, with a module containing a video camera for inspection and mapping, is inserted into a sewer pipe at a manhole. A measuring wheel on the robot's side helps the crew determine the best spots for installing the mounting rings.

2 RING INSTALLATION
When the robot reaches the next manhole, the inspection module is removed and two other modules, one that carries a supply of mounting rings and the other that installs them, are attached. The robot then backs down the pipe, and when it reaches the appropriate spot, mechanical arms grasp a ring from the supply module and a gear expands the ring, much like a hose clamp, until it is flush with the inside of the pipe. The robot then moves on to install the next ring.

3 CONDUIT AND CABLE
On a third pass down the pipe, the robot is equipped with a module to install the conduit. The module includes a laser so that an operator can align the conduit with clamps in the mounting ring. Then the robot pushes the conduit into the clamps. Fiber optic cable is snaked through the conduit later.

ELECTRIC, VIDEO AND PNEUMATIC LINES

ELECTRIC MOTOR

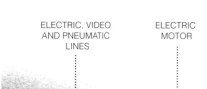

VIDEO CAMERA

ROBOT

MEASURING WHEEL

RING INSTALLATION MODULE

RING BEING INSTALLED

RING SUPPLY MODULE

SEWER PIPE

VIDEO CAMERA

Sources: CityNet Telecommunications Inc., Ka-Te System A.G.

Mika Gröndahl

It's a Dirty Job, but Some Robot's Got to Do It

By John Schwartz

Sitting at the bottom of a manhole in Albuquerque, ready to crawl into a 12-inch sewer pipe, is a long, thin robot on wheels: the Sewer Access Module, or SAM. The robot, which looks a little like the unlikely offspring of a dachshund and an Electrolux vacuum cleaner, was created to fulfill a dream: laying fiber-optic cable within city sewer systems.

Efforts like this promise sweet relief to Americans nationwide who have sacrificed their shock absorbers, wheels and lower backs to streets turned into washboards by companies cutting and repaving streets to lay the "last mile" of fiber-optic cable, the path of choice for high-speed voice and data communications. The owner of the robot, CityNet Telecommunications Inc., asserts that its methods are 60 percent faster than street-cutting methods and that its costs are comparable to those of damaging road work.

The company's chief executive, Robert G. Berger, said the idea of using municipal water systems to deliver bursts of laser light came to him in the shower. His background had prepared him for the epiphany: Mr. Berger, a telecommunications lawyer, was serving his second term on the Washington Suburban Sanitary Commission. "I was looking up at the water, washing the shampoo out of my hair, and said: 'Oh, my gosh! I'm a water and sewer commissioner — it's the one infrastructure that goes everywhere.'"

Mr. Berger is not the first to think of new uses for old city systems. In Chicago, telecommunications companies have taken advantage of abandoned 19th-century tunnels once used to carry coal and ice to downtown buildings. New York has studied whether to run fiber-optic cables through an abandoned network of water mains formerly used by firefighters.

Mr. Berger's ideas had to change somewhat over time: his first idea was to use the water supply system, not the sewers, but the high water pressure, small pipes and safety concerns led him instead to the other end of the water system.

And, inevitably, to robots. Although sewer pipes come in many sizes, few can accommodate a person. Robots, however, come in all shapes and sizes; and when the science fiction visionaries promised that some day robots would handle the dirty tasks that humans would abhor, well, this was exactly the sort of thing they were talking about.

At a trade show, Mr. Berger discovered Ka-Te System A.G., a Swiss company that made sewer-roving robots for maintenance work; it had developed a robot that was laying fiber-optic cable in Hamburg, Germany. CityNet bought the North American rights to SAM's and ordered three.

Robots do not take humans out of the muck entirely, however. Somebody has to get the robot into the manholes, to build in the "slack boxes" that allow the connections from the fiber-optic network into buildings, and to take on other tasks, sometimes unpleasant.

The sewer-going robot is commanded from a truck, where a worker sits at a computer monitor and instrument panel and nudges a nubby rubber joystick forward. The little beast creeps ahead into the pipe on its red rubber wheels, dragging a control cable behind it.

The robot makes at least three trips through each section of pipe. On the first, it carries a camera that records a video of the pipe's interior, to help the workers plan the rest of the operation.

On the second trip, the robot installs stainless steel rings that hold the fiber-optic cable in place (and, as a bonus, reinforce the pipes). Its mapping rig is replaced with a magazine of 10 rings that look a little like an automobile hose clamp; on command at five-foot intervals, the robot slides each ring forward off of the magazine and moves it into position; a gear opens the spring-loaded ring until it sits snugly in place.

On the third visit, the robot carries stainless steel conduit, unwound from a giant spool above ground, and snaps it into place on clamps on the rings. The fiber-optic cable is snaked through the conduit later.

Mr. Berger is not the first to think of new uses for old city systems. In Chicago, telecommunications companies have taken advantage of abandoned 19th-century tunnels once used to carry coal and ice to downtown buildings. New York has studied whether to run fiber-optic cables through an abandoned network of water mains formerly used by firefighters.

The Long and Short of It

Surveyors commonly use a device known as a total station, which has largely replaced equipment like transits and steel measuring tapes. A surveyor holds a prism on a pole at the point being surveyed, and the total station, mounted on a tripod, uses a laser and an electronic measuring device to determine distance and the horizontal and vertical angles to the prism. From this, the location of the point relative to other points can be determined.

VERTICAL DRIVE
A small electric motor drives a large gear that rotates the telescope vertically as a laser tracks the location of the prism.

MOTION CONTROLS
The device can also be rotated manually if the surveyor chooses not to use the laser tracking system but instead lines up the prism by sight, through the telescope.

CIRCUIT BOARD
One of several circuit boards in the device, it handles the distance and angle-measurement data.

ELECTRONIC DISTANCE MEASUREMENT
Light from diodes is directed out through the exact center of the telescope and reflects off the prism back to a detector. By measuring the phase difference between the outgoing and incoming signals, the distance to the prism can be accurately determined.

LASER TRACKING
A laser sends a beam to the prism and, from the strength of the return signal, the drives move horizontally and vertically so the telescope is centered on the prism.

HORIZONTAL DRIVE
A motor drive rotates the device horizontally in response to the laser tracking system.

PLUMBING TELESCOPE
A telescope sights down along the total station's vertical axis, to insure that the device is level and precisely over a fixed point — a benchmark, for example.

DISPLAY
Liquid-crystal screen can display the raw data from each reading (like the distance or vertical angle) or a variety of calculated positions.

Fixing a point

1 Total station determines horizontal and vertical angle from fixed point to prism using laser beam.

2 Electronic measuring device calculates distance from fixed point to prism.

3 Horizontal and vertical angles and distance fix the prism point relative to the total station.

Source: Topcon America Corporation

Frank O'Connell

For Surveyors, Diodes and Lasers Replace Transits and Chains

By Jeffrey Selingo

George Washington, Thomas Jefferson and Abraham Lincoln all worked at one time as surveyors, helping to map thousands of uncharted acres in the land they would one day lead.

While the same principles of distance and angles apply today, modern surveying instruments are smaller, lighter and more accurate than those employed by Washington when he laid out parts of western Virginia. Hand-held computers, lasers and robots have largely replaced the tools once used by surveyors, like the small telescope on a tripod, called the transit, and the 66-foot chain.

"I can't honestly think of a profession more impacted by technology in the last 30 years than surveying," said David Ingram, a surveyor in Virginia and a trustee of the Museum of Surveying in Lansing, Mich. "It's been an exponential leap in capabilities and productivity."

Nearly every construction project or land deal begins with a survey. Surveyors lay out the stakes that mark new highways or housing subdivisions, survey property when it is sold and work for oil and gas companies exploring new energy resources.

In the 18th and 19th centuries, surveying crews sometimes numbering as many as 20 people were sent out for several months to map the West, said John McEntyre, a retired professor emeritus of land surveying at Purdue University. The crews endured rough weather, rugged terrain and confrontations with American Indians. Even today, the sight of a surveyor treading on land can incite property owners, Dr. McEntyre said.

"Surveyors deal with a lot of frayed nerves," he said. "Sometimes it's neighbors who forget where their property lines are or people who don't want a new subdivision built in their backyards. It's always the surveyor who hears about it."

Until about the 1980's, surveying was a cumbersome task, Mr. Ingram said. A typical survey required a crew of three or four people, could take days to complete and was somewhat inaccurate.

Under that system, the crew's operator looked through the transit scope to calculate the angle between a known point (designated by a fixed mark in the ground) and an unknown point. If the ground needed to be flat, the operator also determined where it needed to be leveled by using instruments on the transit similar to those used by carpenters. Those coordinates were then read to another crew member, who recorded the numbers in a book.

Two other crew members held the ends of a 66-foot measuring chain (80 of them made a mile), or a steel tape 50 to 200 feet long, and figured out the distance between the two points. Some surveyors used an iron-nickel alloy tape for extremely precise measurements because those tapes contracted or expanded less with cold or heat.

Still, Mr. Ingram said, the tapes were accurate to within only a tenth of a foot. And a simple task, like measuring a long line, could take days.

Since the 1980's, surveyors have largely employed computerized instruments called total stations to measure distances. The total station uses a laser beam, sent out to a prism on a pole held by a crew member, to bring the total station into precise vertical and horizontal alignment with the prism. Then an electronic measuring device determines the distance to the prism. The location of the pole relative to the total station is then calculated.

The tool allows one or two people to do a surveying job and has increased accuracy to within one-hundredth of a foot, Mr. Ingram said.

Even more recently, surveyors have started to use satellite technology and robots. Under that system, a Global Positioning System receiver is placed at an unknown point and receives a signal from G.P.S. satellites to determine the coordinates of the spot. Even though several cities and counties make surveyors submit measurements using G.P.S. calculations, Mr. Ingram said, the cost of the system has so far prevented its widespread use. The instrument can cost up to $75,000.

Despite the vast changes in technology, some surveyors say it is important for them to know how older surveying instruments worked so they can more easily retrace the steps of their predecessors when doing modern surveys. Dr. McEntyre said many surveyors still carried compasses and chains when researching old property lines.

"Without knowing how the survey was done or with what type of equipment, retracing old surveys and finding old markers is almost impossible," he said. "And no amount of high-tech equipment will help you out if you want to know how a surveyor plotted that land decades ago."

Keeping Watch on the Line

Machine vision is useful in manufacturing and quality control because it can direct robots in specific tasks and detect subtle defects in products in less time than human inspectors. A bottling company, for example, may use machine vision to insure that containers are sound or that labels are affixed correctly.

Most machine-vision applications use a standard 640-by-480-pixel grid for comparing images. Each pixel is one of up to 256 shades of gray, allowing the computer to register more subtle variations between images than the human eye, which can typically distinguish only about 64 shades of gray.

1. A CAMERA photographs a bottle with a defective label.

2. Moments later, the bottle travels by the **REJECT MODULE** and is removed.

DATA OUT

DATA IN

REJECT BIN

DATA IN

A **SENSOR** triggers a frame grabber, which grabs a picture of each bottle as it moves in front of the video camera. Image data are transmitted to a computer where a reference image of what a bottle should look like is stored. Software on the computer then looks for differences in pixel patterns between the reference image and the image of the bottle being inspected. If variations are too extreme, a message is transmitted to the production line to reject the bottle.

Detecting Errors
INSPECTING LABELS

In this case, the computer knows from the reference image that the pixels at or around row 210 should be darker than those in previous rows, indicating the top edge of the label.

To make sure that the label is straight, two points along the label edge are identified. If they are separated by more than a predetermined number of pixel rows, the bottle is rejected.

INSPECTING BOTTLES

Imperfections in a bottle are identified by matching brightness values in the reference image with those in the inspection image. The reference image establishes a pattern of dark and light pixels for the neck of the bottle, for example. Values for the bottle being inspected are subtracted from the values in the inspection image, revealing any defects.

REFERENCE IMAGE
An image of what the area being inspected (in this case, the neck of the bottle) should look like is stored in the computer's memory.

INSPECTION IMAGE
The camera on the assembly line is trained at the same area as the reference image.

COMPARISON
Mathematical formulas are used to compute pixel variations between the images. If there are too many differences, the bottle is discarded.

Another Approach: Finding Just the Right Part

More advanced applications can discern objects, enabling a robot arm to grab parts off a conveyor belt and assemble them. Software searches for patterns in pixels indicating lines and arcs, and notes their relationship. Lines and arcs are compared with the reference image for geometric similarities. Sophisticated mathematical formulas enable the computer to recognize a part and send a signal to pick it up even when it is obscured by shadow or an overlapping object.

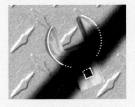

REFERENCE IMAGE IN THE COMPUTER

SHAPES COMPUTER IS LOOKING FOR

ANALYZING THE IMAGE

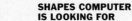

The software looks for sharp differences in pixel values, which suggest edges.

An object can be recognized even if the lighting conditions change or half of it is hidden under another object.

Source: Cognex

Mika Gröndahl and Tom Zeller

Machine Vision Narrows the Gap Between Human and Robot

By James Ryan

If we ever arrive at that point in history where machines become our masters, it will have all begun with machine vision.

Machine vision allows assembly-line robots to locate and identify parts to inspect, count, measure and verify objects in a variety of industrial processes. In other words, for at least some things people can do, they can do them better. "It's a way of giving computers eyes," explained Art Molella, director of the Lemelson Center at the National Museum of American History at the Smithsonian. "It's obviously something that has revolutionized assembly lines."

The essential hardware components are relatively common: a video camera, a personal computer, a sensor and a frame grabber, which captures a still picture from a moving video image.

There is something slightly menacing about hunks of wire and metal that can see more clearly for some tasks, thus performing them with more accuracy than their human creators — even if those hunks of metal are still a far, far cry from the Terminator.

Machine vision was conceived in the mid-1950's by Jerome H. Lemelson, an independent inventor with more than 500 patents.

The Big Three automakers initially rejected Mr. Lemelson's suggestions for automating assembly lines. But Japanese automakers did not, and they leaped ahead in quality and productivity for a while. Detroit eventually caught on. But Mr. Lemelson's contribution to automotive excellence went largely unrewarded until his patents came through decades later.

The importance of machine vision goes beyond its commercial value, Mr. Molella said. "What's more important is the path it has opened toward artificial intelligence and robotics in general," he said. "It has fed the imagination as well as the assembly lines."

Some machine vision systems, like those used to inspect fruit, can "see" in color, but most find black and white adequate. A computer can distinguish between 256 shades of gray. The human eye? About 64 shades.

While humans need two eyes for good depth perception, most machine vision systems find one camera sufficient. Since the system knows the actual size of an object it is scanning for, it can use the size of the object's image and the angle at which light from the object strikes the system's lens to calculate position.

Machine vision has humans beat hands down when it comes to consistency. There's no more worrying that the night shift will see things differently from the day shift. Robots are more precise than assembly line workers, said Joe Campbell, vice president for marketing at Adept Technology, a maker of industrial robots in San Jose, Calif.

A robot with a machine vision system can be built and programmed to perform a wide range of tasks: picking up a single bolt from a jumble on a conveyor belt and screwing it into an engine block, automatically discarding a defective soda bottle in a production line, or placing a needle on the hub of a speedometer moving down a conveyor belt and pointing it exactly to zero. Machine vision systems use video cameras and perceive objects by using mathematical approaches that look for patterns in pixelated video images. Many of the mathematical formulas were developed decades ago.

The biggest advances in recent years have come from improved cable couplings and faster computers, which allow the systems to work more quickly. The speed at which machine vision systems can perform a function is doubling every couple of years, said Tom Reynolds, general manager of Advanced Control Technologies, based in Laguna Hills, Calif.

"Three years ago, we could do 3 inspections a second," Mr. Reynolds said. "Now we can do 10; a year from now it will be 20."

Let's see, that means we humans should be safe from machines' taking over the world until, oh, at least 2033.

You Look Familiar

The use of biometrics for facial recognition is making it possible for people to be identified without presenting passwords or PIN's. The different technologies convert physical characteristics into patterns or mathematical renderings that can be analyzed by computers.

▶ Iris Scans

The tangled mesh of connective tissue in the iris is unique for each eye. It is stable throughout life, making it an ideal feature for biometric identification.

IriScan

SCANNING THE EYE

The iris pattern is encoded into an Iris Code, which is stored in a database. The error rate for identification is 1 in 1.2 million.

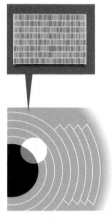

IRIS CODE

▶ Eigenfaces

1 A large number of pictures of faces are collected in a database.

2 A set of eigenfaces — two-dimensional facelike arrangements of light and dark areas — is made by combining all the pictures and looking at what is common to groups of individuals and where they differ most.

M.I.T. Media Laboratory

EIGENFACES

3 Eigenfaces work as "primary faces." Just as any color can be created by mixing primary colors, any facial image can be built by adding together, with different intensities of light, a relatively small number of eigenfaces (about 100 is enough to identify virtually any person).

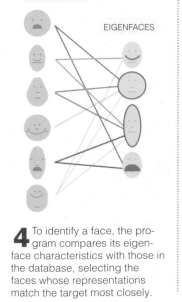

REAL FACES IN DATABASE

EIGENFACES

4 To identify a face, the program compares its eigenface characteristics with those in the database, selecting the faces whose representations match the target most closely.

▶ Local Feature Analysis

1 Local feature analysis is derived from the eigenface method but overcomes some of its problems by not being sensitive to deformations in the face and changes in poses and lighting. It considers individual features instead of relying on only a global representation of the face. The system selects a series of blocks that best define an individual face from which all facial images can be constructed.

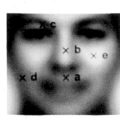

A SMALL NUMBER OF FEATURES MAKE EACH FACE UNIQUE.

THE MOST-DEFINING POINTS OF THE SAME INDIVIDUAL, HIGHLIGHTED.

2 The procedure starts with eigenfaces from a database of photos.

3 Applying local feature analysis, the system selects the subset of building blocks, or features, in each face that differ most from other faces. The most characteristic points are the nose, eyebrows, mouth and the areas where the curvature of the bones changes.

4 The patterns have to be elastic to describe possible movements or changes of expression. The computer knows that those points can move slightly across the face in combination with the others without losing the basic structure that defines that face.

Visionics

5 To determine someone's identity, **(a)** the computer takes an image of that person and **(b)** determines the pattern of points that make that individual differ most from other people. Then the system starts creating patterns, **(c)** either randomly or **(d)** based on the average eigenface. For each selection, **(e)** it constructs a facial image and compares it with the target face to be identified. **(f)** New patterns are created until **(g)** a facial image that matches the target can be constructed. When a match is found, the computer looks in its database for a matching pattern of a real person **(h)**.

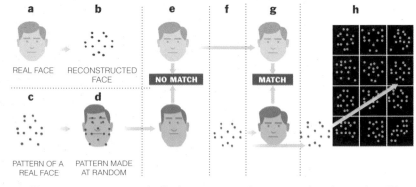

a REAL FACE
b RECONSTRUCTED FACE
e NO MATCH
f
g MATCH
h

c PATTERN OF A REAL FACE
d PATTERN MADE AT RANDOM

6 The system can automatically detect a person's presence, locating and tracking the head. It can work from any distance (depending on the optics) and even recognize moving people.

Sources: Visionics Corporation; Miros, Inc.; IriScan Company; Unisys Corporation; Mr. Payroll Corporation; M.I.T. Media Laboratory; Peter Kruizinga (University of Groningen)

Juan Velasco

Teaching the Computer to Recognize a Friendly Face

By Juan Velasco

Are you tired of remembering so many passwords? Would you like to get rid of your PIN for the cash machines and those annoying passwords you use to log on to different Web sites and the office computer? How about trading them for a simple smile?

That is the promise of facial recognition, an aspect of personal-identification biometrics technology that once seemed the province of James Bond movies. Biometrics measures the physical characteristics that make each person unique, like the fingerprints and voice or the distinctive, complex patterns in the eye's iris or retina.

People might find it hard to remember faces, but machines that use facial recognition technology are not likely to forget one. They can even spot a face in a crowd. Facial recognition could be used to control access to restricted areas in corporate offices or airports, to identify people crossing borders or to maintain privacy for medical records and financial transactions at A.T.M.'s.

The technology requires little or no cooperation from the subject and, unlike other biometric techniques, it can be easily backed up by people without special expertise — anyone can recognize a face.

Most facial recognition systems work in the same basic way. The machine takes an image of the face obtained with a video camera and identifies it by matching it against a database of faces. Systems differ in how they obtain the images — the most advanced systems automatically detect any human presence, locating and tracking the person's head, and can work under different lighting conditions and on moving subjects, accommodating changes in expressions or hairstyles.

In one system, used to identify people trying to cash checks at A.T.M.'s, customers who use the machine for the first time are photographed and asked a series of verifying questions. After that, all they need are their Social Security numbers and their presence.

The system, developed by Miros Inc. of Wellesley, Mass., is based on a computer with artificial neural networks, which are designed to process information in much the way the brain does. The greater the number of pictures of people in the database, the faster the machine can recognize people. It works by looking at facial features like the cheekbones and jaw lines.

Another system, developed by Visionics Corp., uses a mathematical method called local feature analysis and combines it with the eigenheads approach, derived from an earlier method called eigenfaces, invented in 1987. But unlike the eigenfaces or artificial neural networks techniques, which focus on a global representation of the face, the Visionics system considers individual features, like the nose, mouth or eyes, and can identify a moving face.

Another type of face recognition focuses on the iris of the eye. Iris recognition technology is based on the fact that the random patterns found in irises are unique, the equivalent of human bar codes. In the entire human population, no two irises are alike. A person's iris pattern is encoded into an Iris Code, which is stored in a database and used for recognition when a live iris is presented for comparison. Glasses and ordinary contact lenses do not interfere with recognition.

Finally there are facial thermograms, an experimental system that uses an infrared camera to look at the heat emission patterns on each face. The map that results is different for each person.

The Road to the Bar Code

The United States Postal Service automates mail sorting by translating ZIP codes into bar codes, which can be scanned by machines. Because mail comes in so many varieties, however, insuring that each piece of mail gets the proper bar code becomes a complicated task. The postal service employs digital imaging and optical character recognition to sort, cancel and deliver more than 600 million pieces of mail per day.

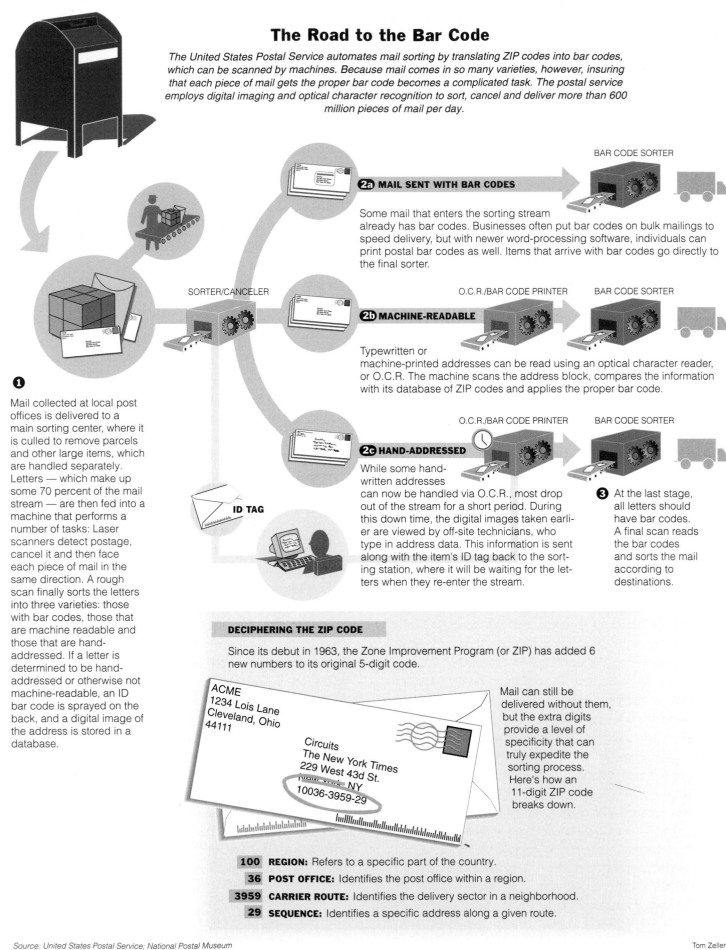

BAR CODE SORTER

2a MAIL SENT WITH BAR CODES

Some mail that enters the sorting stream already has bar codes. Businesses often put bar codes on bulk mailings to speed delivery, but with newer word-processing software, individuals can print postal bar codes as well. Items that arrive with bar codes go directly to the final sorter.

SORTER/CANCELER

O.C.R./BAR CODE PRINTER   BAR CODE SORTER

2b MACHINE-READABLE

Typewritten or machine-printed addresses can be read using an optical character reader, or O.C.R. The machine scans the address block, compares the information with its database of ZIP codes and applies the proper bar code.

O.C.R./BAR CODE PRINTER   BAR CODE SORTER

2c HAND-ADDRESSED

While some hand-written addresses can now be handled via O.C.R., most drop out of the stream for a short period. During this down time, the digital images taken earlier are viewed by off-site technicians, who type in address data. This information is sent along with the item's ID tag back to the sorting station, where it will be waiting for the letters when they re-enter the stream.

❸ At the last stage, all letters should have bar codes. A final scan reads the bar codes and sorts the mail according to destinations.

ID TAG

❶

Mail collected at local post offices is delivered to a main sorting center, where it is culled to remove parcels and other large items, which are handled separately. Letters — which make up some 70 percent of the mail stream — are then fed into a machine that performs a number of tasks: Laser scanners detect postage, cancel it and then face each piece of mail in the same direction. A rough scan finally sorts the letters into three varieties: those with bar codes, those that are machine readable and those that are hand-addressed. If a letter is determined to be hand-addressed or otherwise not machine-readable, an ID bar code is sprayed on the back, and a digital image of the address is stored in a database.

DECIPHERING THE ZIP CODE

Since its debut in 1963, the Zone Improvement Program (or ZIP) has added 6 new numbers to its original 5-digit code.

ACME
1234 Lois Lane
Cleveland, Ohio
44111

Circuits
The New York Times
229 West 43d St.
NEW YORK, NY
10036-3959-29

Mail can still be delivered without them, but the extra digits provide a level of specificity that can truly expedite the sorting process. Here's how an 11-digit ZIP code breaks down.

100 REGION: Refers to a specific part of the country.

36 POST OFFICE: Identifies the post office within a region.

3959 CARRIER ROUTE: Identifies the delivery sector in a neighborhood.

29 SEQUENCE: Identifies a specific address along a given route.

Source: United States Postal Service; National Postal Museum

Tom Zeller

It May Be Snail Mail, but Technology Helps Get It Where It's Going

By Peter Wayner

If there is one institution that would seem to be threatened by the information superhighway, it's the United States Postal Service. The world of e-mail, Web pages and software downloads seems to offer everything the old paper-based world of information delivers, but offers it almost instantaneously. Can something known colloquially as snail mail compete?

The answer is that it won't compete directly, but it will keep applying the same kind of computer power that underlies e-mail and other electronic wizardry to the task of moving paper mail. The fact is that new technology rarely destroys the old. New and old usually meld, allowing both to do what they do best.

Letter carriers deliver hundreds of millions of pieces of mail a day. The post office now uses computerized sorting machines to route most of the mail. Most bulk mail is printed with bar codes that are representations of ZIP codes and easy for computers to understand. The machines direct the path of a letter with house-level precision.

Mail with bar codes goes directly to machines that can read the codes. Mail without bar codes but with printed addresses is handled by large mail-sorting machines with built-in cameras for taking pictures of addresses. A computer interprets the images, finds the best matching addresses and directs a printer to mark bar codes on the envelopes to speed further sorting.

The post office is concentrating on building even more sophisticated computer systems, with advanced optical character recognition (O.C.R.) technology, that will read handwritten envelopes. Bill Dowling, a vice president for engineering at the Postal Service, said, "Our online O.C.R. machines handle 40,000 envelopes an hour and 80 percent of machine-printed mail we read."

When letters without bar codes go into the first sorting machine, the computers try to interpret the address in the fraction of a second between when the letter passes by the camera and when it passes by an ink-jet printer that sprays on bar codes in fluorescent ink. Finding the answer in this short amount of time is called online processing.

The original computer machines doing this job relied upon VAX minicomputers, which were manufactured by the old Digital Equipment computer company that is now a division of Compaq. Each machine is now equipped with additional chips that analyze the address images together. These extra chips run a stripped-down version of Linux that the Postal Service customized for high-speed parallel processing of addresses.

Hundreds of O.C.R. machines are distributed throughout the country at different sorting centers. Many smaller branches and local post offices have cheaper, faster sorting devices that can read only bar codes.

If the computer cannot recognize the address in the short amount of time available to it — perhaps because the address is handwritten — the letter and the digital image of the address are put aside for off-line processing. First, a computer tries to interpret the image with more sophisticated algorithms (mathematical approaches) that take several seconds to search for an answer. If these computers cannot read the address, the image of the address is finally given to a person, who reads it and types it into the system.

Then the batch of envelopes with addresses that were hard to read are processed again. This time, the correct bar code is sprayed on each envelope so it can be sorted at local post offices.

Both the printed-character and handwritten-character recognition packages rely heavily on a large database containing every address in the country.

"We have an excellent digit recognizer so we can recognize the ZIP code and correlate the city and state," said Ed Kupert, a manager of image-handling technology for the Postal Service. "Then we can recognize the house number. From this, we can create a lexicon of every street that has a house with that number in the ZIP code, and the computer picks the best match."

The Postal Service is also following the lead of other delivery companies and using technology to add new features to the seemingly simple process of delivering a letter. One option would use special bar codes to track the flow of individual envelopes in much the same way that package delivery businesses like Federal Express and United Parcel Service follow packages. Credit-card companies and mortgage holders may want to be able to know if the payment check is really in the mail — and if so, when it got there, Mr. Dowling said.

Putting Bills to the Test

There can hardly be a consumer who hasn't had many experiences with the bill validators inside soda and candy machines or the ones that scrutinize a $5 bill when change is needed for the laundry. These machines have microprocessors and optical sensing devices that make more trouble for counterfeiters and less for consumers.

Jack Manning

STACKER PLATE

CIRCUIT BOARD

ELECTRIC MOTORS FOR DRIVE BELTS AND STACKER

Bills are pulled in by two rubber rollers and belts, then stacked in a magazine by a stacker plate.

Currency acceptors or validators use optical sensors to determine if it is real. A roller mechanism is triggered when a bill is detected by a sensor beam and pulls the bill past a group of vertical sensors.

BILL STORAGE

DRIVE BELTS

SENSOR UNIT

INSIDE THE SENSOR UNIT

The sensor unit contains four vertical sensors (grouped in pairs), and each sensor has two parts. Below the bill are small cylinders, each about a quarter-inch across, containing light-emitting diodes that shine light upward, through the bill. Above the bill, receptors measure how much light passes through. The gap between the L.E.D.'s and the receptors — the bill path — is about one-tenth of an inch.

Each **SENSOR** measures a quarter-inch swath.

Light hitting the receptors, or photo diodes, is translated into **CURRENT.**

The current is translated into **DIGITAL INFORMATION.**

The microprocessor compares stored data describing what a certain spot on the bill should look like (that is, how much light should pass through) with the data that the optical sensors provide. Among the things that affect this are paper quality and thickness, ink quality and the special threads that are embedded in bills to deter counterfeiters. The data also tell the microprocessor what kind of bill it is. If change is necessary, a signal is sent to that processing unit.

ROLLERS

VERTICAL SENSORS
Receptors
Light beam
Light-emitting diodes

ROLLER ACTIVATING BEAM

LIGHT-EMITTING DIODES

Each vertical sensor can contain several different diodes, and each diode can emit a different frequency of light in the visible or infrared spectrum.

HORIZONTAL LIGHT BEAM

To thwart thieves who attach something to a bill so it can be pulled out after it is scanned, a horizontal light beam is used to determine whether anything is stuck to the bill.

Source: Mars Electronics International

Mika Gröndahl

In Vending Machines, Brains That Tell Good Money From Bad

By James Ryan

If you feed a dollar into a machine and it obstinately spits it back at you — no matter how carefully you smooth away the tiny folds and wrinkles — chances are you're nowhere near a Las Vegas casino. In their ongoing efforts to separate fools from their money, casino operators have always been among the first to adopt the latest bill validator technology.

Elsewhere, it may seem at times as if the bill acceptors in vending and change machines were devised by the Treasury gods to torture mere mortals. In fact, the devices exist to reject bogus bills.

Bill acceptors were developed in the late 1960's primarily to make it easier to sell sodas and candy bars from vending machines. It's no accident that one of the leading suppliers of the devices, Mars Electronics International, began life as a division of the Mars candy company.

The first generation of machines used a magnetic head, like those in audio cassette players, to read the ink on a dollar bill. But the magnetic heads needed to be in contact with the bill and would frequently become fouled with grime and lint. The process was possible because the ink used by the Treasury has a high iron content — but so does the ink in some copy machines. As recently as the early 1990's, it was relatively easy to trick a dollar machine with a black-and-white photocopy.

"People fool bill validators with things that would not fool humans," explained Tom Shuren, director of engineering at Mars Electronics International. Newer magnetic readers float above the surface of the bill and can distinguish between copier ink and Treasury ink by measuring not just the magnetic field but the pattern of magnetic particles within the ink, which serve as a sort of fingerprint.

Another way of fooling bill readers involved attaching a string or piece of tape to the back of a bill, enabling a thief to pull it out of the machine once it was recognized.

"Some of these thieves could unload $500 from a changer in a matter of minutes," said Tom Reynolds, president of Cointrol, a major distributor of dollar changers. Most devices today have a sensor that looks for foreign objects attached to the back of a bill.

To foil sophisticated thieves, in the early 90's manufactur-ers of bill acceptors began using optical sensors, sometimes in tandem with magnetic sensors. The optical sensors use microprocessors to analyze the shadows cast by various parts of a bill. Each sensor focuses on specific spots in a narrow swath of the bill as it passes by. (Precisely which parts of the bill are looked at cannot be revealed so as not to abet counterfeiters.)

Some of the most advanced machines come from Cash Code, a company in Concord, Ontario. Among the employees giving Cash Code a competitive edge are a group of Russian émigré engineers who once worked on the Soviet missile and space programs.

"It's a never-ending game — as we develop new methods of detecting fakes, counterfeiters develop new methods," said Val Levitan, senior vice president of marketing at Cash Code.

Mr. Levitan said Cash Code's newer devices employ capacitive, or dielectric, sensors, which measure conductivity and can detect phony watermarks or fake identifying threads in counterfeit bills.

The more recent models of bill validators are also much more flexible. When the new $20 bills came out, many older machines would not accept the bills and had to be scrapped or undergo costly retrofitting, with the installation of new chips. Newer machines, however, can be "flash" upgraded to accommodate new bill designs: a technician merely downloads the new software program from a laptop.

To make sure that an advanced bill validator can read slightly tattered, faded or soiled bills, employees artificially age bills by wrinkling them, passing them hand to hand and even running them through a washing machine. Machines are then calibrated to accept these less-than-perfect bills. The industry standard is an acceptance rate of 96 percent. If your corner Pepsi machine appears to have a much lower rate, it is probably an older machine or is not serviced regularly. "You don't experience as many problems when you are in Vegas because they have service technicians on the spot," Mr. Levitan said. "They have those machines up and running in minutes."

Show Me the Money

The automatic teller machine, first introduced in the United States at a bank in Georgia, has become as ubiquitous as the gas pump. There are now tens of thousands of A.T.M.s in the United States, at banks but also at supermarkets, convenience stores, highway rest stops, sports arenas — in short, anywhere people congregate and spend money. Here's a look at how a typical A.T.M. works.

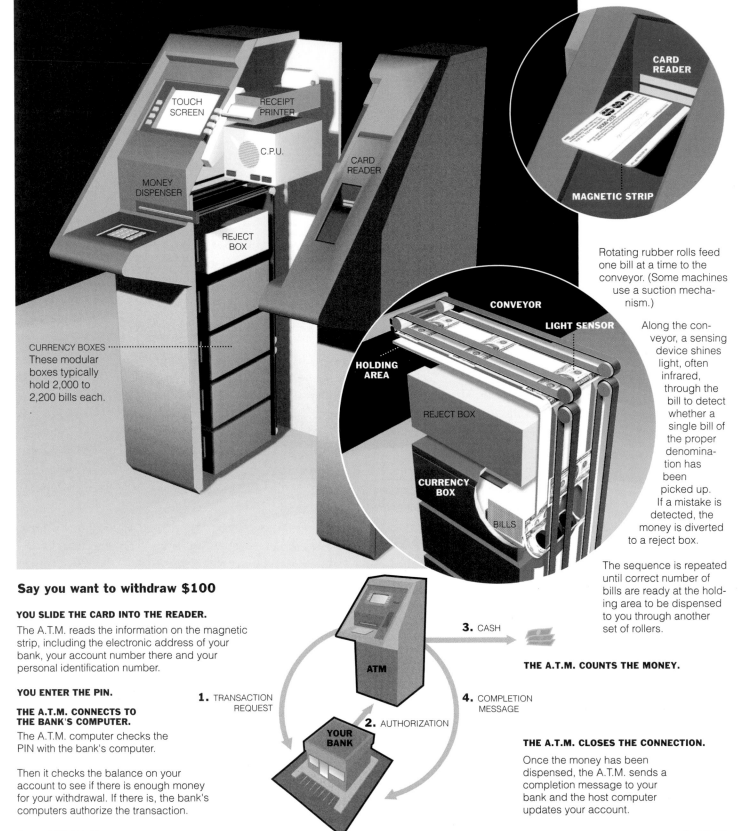

Some card strips carry personalized information so the A.T.M. can greet you by name: Welcome, Bob Smith.

CARD READER

MAGNETIC STRIP

TOUCH SCREEN

RECEIPT PRINTER

C.P.U.

MONEY DISPENSER

CARD READER

REJECT BOX

CURRENCY BOXES
These modular boxes typically hold 2,000 to 2,200 bills each.

CONVEYOR

LIGHT SENSOR

HOLDING AREA

REJECT BOX

CURRENCY BOX

BILLS

Rotating rubber rolls feed one bill at a time to the conveyor. (Some machines use a suction mechanism.)

Along the conveyor, a sensing device shines light, often infrared, through the bill to detect whether a single bill of the proper denomination has been picked up. If a mistake is detected, the money is diverted to a reject box.

The sequence is repeated until correct number of bills are ready at the holding area to be dispensed to you through another set of rollers.

Say you want to withdraw $100

YOU SLIDE THE CARD INTO THE READER.

The A.T.M. reads the information on the magnetic strip, including the electronic address of your bank, your account number there and your personal identification number.

YOU ENTER THE PIN.

THE A.T.M. CONNECTS TO THE BANK'S COMPUTER.

The A.T.M. computer checks the PIN with the bank's computer.

Then it checks the balance on your account to see if there is enough money for your withdrawal. If there is, the bank's computers authorize the transaction.

ATM

1. TRANSACTION REQUEST

2. AUTHORIZATION

YOUR BANK

3. CASH

4. COMPLETION MESSAGE

THE A.T.M. COUNTS THE MONEY.

THE A.T.M. CLOSES THE CONNECTION.

Once the money has been dispensed, the A.T.M. sends a completion message to your bank and the host computer updates your account.

Sources: NCR; Diebold

Mika Gröndahl

Always Convenient, Superbly Efficient: The A.T.M.

By Henry Fountain

It's a fantasy for modern times: You ask an automated teller machine for $100 from your account and like a slot machine showing triple cherries, it just keeps spitting out money.

Alas, such a scene is extremely unlikely, A.T.M. manufacturers say. The machines contain a series of safeguards to make sure that only the exact amount of money requested is dispensed. In essence, said Thomas J. McBride, director of worldwide product marketing for Diebold, a major A.T.M. manufacturer based in Canton, Ohio, the machines test and retest the bills during each transaction in much the way that a human teller counts and recounts money several times before handing it over.

One common testing method uses infrared light. After a bill is selected from a supply hopper, it passes over a beam of infrared light. Varying amounts of light will pass through different parts of a bill to be picked up by a sensor on the other side. Each denomination has a printing signature that the sensor can be programmed to detect. The portrait of Andrew Jackson on a $20 bill, for example, will be read differently than that of Alexander Hamilton on a $10 bill. The sensor can also easily detect bills that are stuck together.

A.T.M.'s also have electromechanical devices that conduct similar tests. These are usually a series of pivoting rollers that check each bill's thickness and size and whether it is skewed and likely to jam the machine. All the devices, whether optical or electromechanical, are calibrated to accommodate anything from a crisp new bill to a worn one.

A bill that fails any test is diverted to a reject hopper, and another bill is drawn from the supply hopper to be run through the testing sequence. Only when the correct number of bills of the right denomination have passed the tests is the money actually dispensed.

Rather than issuing too much cash, what is more likely to happen — though it is still rare, A.T.M. manufacturers say — is that a machine will dispense too little. You might ask for $100, for example, but something will jam inside the machine and only $80 will be dispensed. In such a case, only $80 will be deducted from your account because the electronic transaction with your bank takes place only after the money is given out.

A far different problem arises when an A.T.M. hands over the right amount of money to the wrong person. A personal identification number is the chief means by which an A.T.M. knows that a card user is legitimate, and both card and PIN can be stolen.

Manufacturers are working to increase security, testing verification technologies like voice and face recognition as well as iris and fingerprint scanning. But privacy issues that would be raised by any effort to create a large, centralized database of scans or other information are likely to keep such projects limited.

A different kind of digital safeguard has some potential, said Bill Witte, vice president for product management at Fujitsu-ICL Systems, an A.T.M. manufacturer in La Jolla, Calif. With all the advances in storing computer data, A.T.M.'s can now keep electronic, rather than paper, records of all transactions. It would be simple to add to each transaction record an image of the customer taken by a digital camera on the front of the machine. While such a system might not stop a thief in the act, it could make it easier to track one down later.

What Goes Up Must Come Down, Safely

Advanced elevator systems use computers to monitor and control the movement of cars so that waits are shorter and rides are quicker.

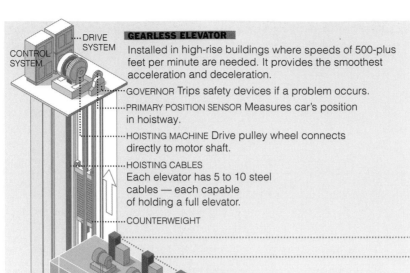

CONTROL SYSTEM

DRIVE SYSTEM

GEARLESS ELEVATOR
Installed in high-rise buildings where speeds of 500-plus feet per minute are needed. It provides the smoothest acceleration and deceleration.

GOVERNOR Trips safety devices if a problem occurs.

PRIMARY POSITION SENSOR Measures car's position in hoistway.

HOISTING MACHINE Drive pulley wheel connects directly to motor shaft.

HOISTING CABLES
Each elevator has 5 to 10 steel cables — each capable of holding a full elevator.

COUNTERWEIGHT

GUIDE RAILS

ROLLER GUIDES

SECONDARY POSITION SENSOR
Measures car's position in hoistway.

DOOR OPERATOR

LOAD-WEIGHING SENSORS
Measure car's load.

CAR SAFETY DEVICES Grab guide rails if programmed speed is exceeded, preventing car from falling.

TRAVELING CABLE Connects car to electrical systems.

COMPENSATION CABLES Keep the system in balance, no matter where the car is positioned.

BUFFER Stops car if it passes lowest doorway.

CONTROLLED BY FUZZY LOGIC
Every time a hall button is pushed, the elevator system's software is set in motion, performing split-second calculations, based on 21 variables, to dispatch the best car. The system can respond to demand based on what it has learned about traffic patterns.

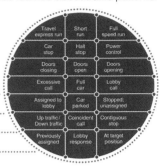

Travel express run	Short run	Full speed run
Car stop	Hall stop	Power control
Doors closing	Doors open	Doors opening
Excessive call	Full car	Lobby call
Assigned to lobby	Car parked	Stopped; unassigned
Up traffic/ Down traffic	Coincident call	Contiguous stop
Previously assigned	Lobby response	At target position

The building manager monitors the system's operation on a computer. Animations show which cars are moving, whether doors are open and other system information.

Operations are monitored by Otis's remote maintenance system. If problems appear imminent, a technician is dispatched, ideally before breakdowns occur.

6 After weighing this car, which is traveling light, the system tells it to stop on the 8th floor.

5 The system determines that the closest car should not stop on the 8th floor because it is near its weight limit.

4 A person on the 8th floor pushes the Up button, seeking to go to the 9th floor.

3 The system decides an empty car is the best one to send to the ground floor, even though it is farther away than others.

2 Early car announcement system instructs the light outside the door of the arriving car to illuminate immediately. That allows passengers ample time to get to the doors, lessening the loading time and allowing the car to leave more quickly.

1 People enter the office building and push the Up elevator button. The system determines which elevator can arrive most quickly, based on how many other stops it is making to pick up and discharge passengers.

ENTRANCE PROTECTION
Fifty-six infrared beams create an invisible net across the elevator entrance. If any beam is interrupted, the system reopens the door instantly — without touching passengers. Additional beams project into the hallway to detect passengers as they enter or exit the elevator. That helps avoid a potential collision between slow-moving passengers and the door. Elevators are also routed if possible to avoid entering and exiting jams, particularly on the ground floor.

INFRARED BEAMS

Source: Otis Elevator Company

John Grimwade

Advanced Elevators: Mass Transit of the Vertical Variety

By Eric A. Taub

Bleary-eyed and coffee-saturated, you wait for the commuter train in the morning, then the subway, then for 4,000 people to cross 53rd Street. By the time you force yourself through the revolving doors to get to work, the last thing you want to do is wait for a crowded elevator to haul you up to your office.

We could all take a lesson from the Three Stooges, for they knew how to get an elevator pronto: a simple yank on the floor indicator stick above the door inevitably brought the car to them instantaneously — along with a jumble of bewildered passengers picking themselves up off the floor.

That technique doesn't work that well today. Among other things, mechanical floor indicators have all but disappeared. In the last 50 years, with the banishment of human operators, elevator management has become a computer-controlled science.

The Holy Grail, according to Bruce A. Powell, perhaps the industry's only certified elevator scholar, is to make elevator operation seem effortless from the point of view of passengers.

As sole possessor of the title "technical fellow" at Otis Elevator Company in Farmington, Conn., the industry leader, Mr. Powell directs a team of people trying to figure out how to move as many people as possible as quickly as possible in an elevator.

Mechanically, elevators use the same basic design they did when Elisha Otis invented the safety version in the 19th century (the first modern elevator is said to have been built in 1743 in France as a way for Louis XV or his mistress to slink around the Palace of Versailles undetected). Cars ride up and down a vertical track, raised by either a hydraulic system or a counterweight. A series of cables and ratchets stop the cab from falling; if a cab does fall, a safety brake kicks in to slow and stop it.

Redundant systems allow elevators to travel at high speeds safely. The world's fastest elevators can move at nearly 30 miles per hour.

In the 1950's, automatic elevators went to floors in the order the call buttons were pushed. Like an emotionally dependent golden retriever, a car would go to the floor that issued the first call, leaving other would-be passengers stranded on distant floors. Elevators would arrive at lobbies in packs, often empty.

By the beginning of the next decade, mechanical elevator dispatchers had been programmed to send cars all the way up and all the way down, stopping for passengers along the way — great if you were the first one on or off, but frustrating if you were along for the long ride. With zoned dispatching, introduced in the mid-1960's, elevators were designated as expresses, serving just lower or higher floors, or as locals. Zone system technology is still used in many American office buildings.

To make elevators more efficient, the industry developed a management system that makes use of fuzzy logic, which allows systems to keep several objectives in mind as it juggles elevator routes. If someone on the eighth floor calls for the elevator, the electronic controller is told, in effect: "Hey, stupid, don't just send the closest cab. Send the one a few floors farther away if it's empty because it'll actually get there faster."

First used in a building in Osaka, Japan, in 1993, fuzzy logic came to American elevators several years later in an Otis retrofit of the systems in the TransAmerica Building in San Francisco. This is fuzzy logic with a twist: not only did TransAmerica workers save time in their vertical journeys, but they also found out which elevator they would be taking the moment they asked because of Otis's early car announcement technology. When a passenger asks for an elevator, the system determines which car the passenger should take and directs the passenger to that elevator door immediately by flashing a light over that door — instead of indicating the car shortly before its arrival.

Yet the system can be misunderstood. After the system was installed at TransAmerica, an angry executive, misunderstanding the advance signal, complained to the building manager that the elevators must be broken because "the elevator light came on and no elevator came for a long time." Lunchtime lectures were quickly held to explain the new system.

Letting It Flow

Automatic faucets have been installed in many public restrooms to conserve water and improve sanitation. Most make use of infrared technology, and some use fiber optics as well. Here's how a typical fiber-optic model knows when to turn the water on.

FAUCET HEAD

INFRARED LENSES

INFRARED EYES

Dual fiber-optic lines inside the faucet head work together to sense hands in the sink. One line emits beams of infrared light, while the other receives the beams reflected back off the hands. A sensor at the end of the receiver line is calibrated to detect light reflected at a certain intensity so reflections from passers-by or from the bottom of the sink are ignored.

FIBER OPTICS

- OUTGOING BEAM
- REFLECTED BEAM
- IGNORED REFLECTION
- FAUCET LINE

SOLENOID VALVE

AUTOMATIC FAUCET UNIT

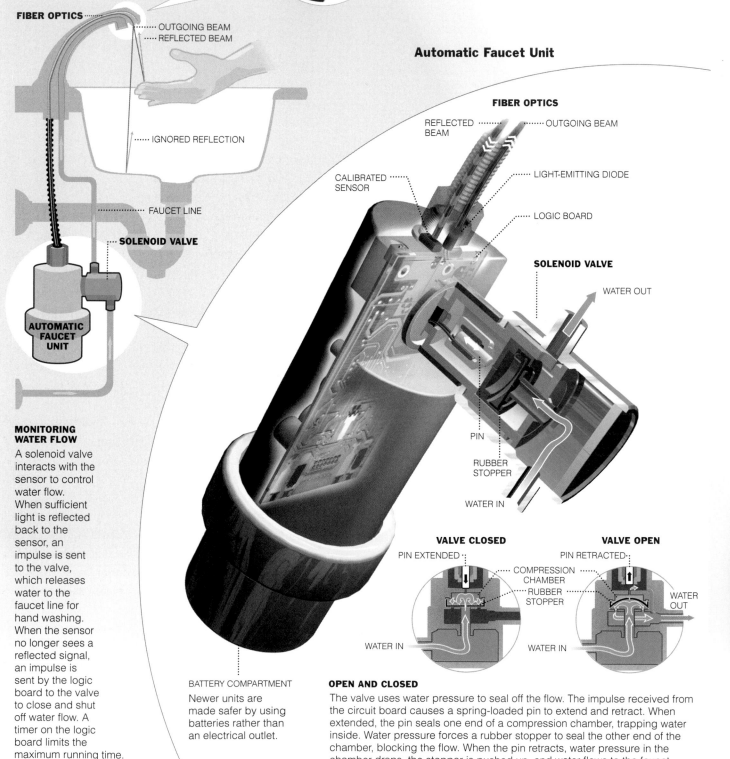

Automatic Faucet Unit

FIBER OPTICS

- REFLECTED BEAM
- OUTGOING BEAM
- CALIBRATED SENSOR
- LIGHT-EMITTING DIODE
- LOGIC BOARD

SOLENOID VALVE

WATER OUT

PIN

RUBBER STOPPER

WATER IN

MONITORING WATER FLOW

A solenoid valve interacts with the sensor to control water flow. When sufficient light is reflected back to the sensor, an impulse is sent to the valve, which releases water to the faucet line for hand washing. When the sensor no longer sees a reflected signal, an impulse is sent by the logic board to the valve to close and shut off water flow. A timer on the logic board limits the maximum running time.

BATTERY COMPARTMENT
Newer units are made safer by using batteries rather than an electrical outlet.

VALVE CLOSED
PIN EXTENDED

COMPRESSION CHAMBER
RUBBER STOPPER

WATER IN

VALVE OPEN
PIN RETRACTED

WATER OUT

WATER IN

OPEN AND CLOSED

The valve uses water pressure to seal off the flow. The impulse received from the circuit board causes a spring-loaded pin to extend and retract. When extended, the pin seals one end of a compression chamber, trapping water inside. Water pressure forces a rubber stopper to seal the other end of the chamber, blocking the flow. When the pin retracts, water pressure in the chamber drops, the stopper is pushed up, and water flows to the faucet.

Source: Sloan Valve Company

Mika Gröndahl and Tom Zeller

Smart Plumbing Never Leaves the Water Running

By Tina Kelley

Sinks, toilets and urinals in many public restrooms use infrared sensors to detect the presence of a person and activate the plumbing. There's no creepy voyeur lurking behind that small, dark sensor panel, just a lot of circuitry.

On the wall or on the faucet, depending on the fixture, there is a black window about an inch long. Behind it is a light-emitting diode that is about the diameter of a ballpoint pen and gives off infrared light. It has a wavelength of 850 nanometers (850 billionths of a meter) and is not visible to the human eye, which can see only the spectrum from violet to red, ranging from 350 or 400 nanometers to 750 or 800 nanometers.

The system is designed to save water by keeping people from leaving faucets on and flushing too often, and to reduce the spread of germs by eliminating faucet and flush handles. Because automatic faucets and fixtures are touched less often, they also need fewer repairs and less scrubbing, which lowers maintenance costs, plumbing manufacturers say.

But automatic fixtures can also make life more complicated. Especially when it comes to toilets.

With the automatic toilets at the new electrical engineering building at the University of Washington in Seattle, "it's a riot," said Martin A. Afromowitz, a professor. "You sit down and sit forward or back, and if your head's down, it flushes. We get triple-flushed and quadruple-flushed. It can give you a heart attack when you don't expect it."

Other variables can also fake out the automatic-flushing system.

"Eighty percent of women hover," said Steve Bissell, senior product manager for the Kohler Company in Kohler, Wis. "They never sit on the fixture, or they sit much farther from the sensor. But the average man is the one who sets up the sensor." Such confusion can keep a toilet from flushing on its own.

Kohler has developed a system to keep the toilet from flushing at the wrong time. It programs the sensor to wait for three seconds after the end of the reflected-light signal indicating that the toilet is in use before flushing.

"Children are a problem because they can be very tiny," said John Lauer, manager of installation engineering at Sloan Valve Company, in Franklin Park, Ill. "We always have on toilets a manual override button. It gives the person a little bit of personal control."

Infrared sensors may react differently to different colors because lighter shades reflect more light. But that was more of a problem in the early 70's, when sensors were less precise and underreacted to dark colors.

While sensors generally seem to have worked well enough to have gained broad acceptance, there are still a number of factors that can cause malfunctions. Sensor function can be affected by a change in the amount of light in a room, perhaps caused by dying fluorescent light bulbs.

"Touchless faucets" are not cheap — they range in price from $300 to $600. The assumption is that they will reduce water use, perhaps by up to 50 percent, even though they also consume power for operation. Most of them use AA or AAA batteries, which can last for two years, manufacturers say.

Infrared plumbing technology has been around longer than one might expect. It has been a big hit at the Madonna Inn in San Luis Obispo, Calif., since it opened in the late 1950's. The inn has what it calls a "world famous" men's bathroom and "caveman" room. Both feature urinals made of heavy rock formations. When motion is detected in the room, water runs down the rocks, ready to serve.

It's All There, in Black and White

Bar codes are composed of a series of parallel black and white lines. The black modules absorb light from the scanner, while the white modules reflect it. A photo diode in the scanner decodes the reflected light into an electrical signal, which is translated into digital code by the scanner's microprocessor.

LEFT HALF

QUIET ZONE
A typical U.P.C., read from left to right, starts with a quiet zone (nine white modules).

GUARD PATTERN
Next comes a three-module guard pattern, which alerts the scanner and computer system that information is coming.

CATEGORY
The next seven modules identify the number of the product category. That number is printed at the lower left margin of the U.P.C.

MANUFACTURER
The next five digits (35 modules) identify the manufacturer.

CENTER GUARD PATTERN
A five-module pattern divides the left and right halves of the bar code.

RIGHT HALF

PRODUCT IDENTIFICATION
The next five digits (35 modules) on the right side identify the product — that is, whether it is a 12-ounce jar of Brand X chunky peanut butter or an 8-ounce jar of smooth.

CHECK DIGIT
The check digit (seven modules) is printed in the lower right margin of the U.P.C.

GUARD PATTERN
Three modules.

QUIET ZONE
Nine modules.

MODULE

MODULE

Pattern for 6 on the right side of the bar code

`1 0 1 0 0 0 0`

0 12345 67890 5

TWO SIDES TO EVERY CODE

In this type of U.P.C., seven modules in the bar code represent one digit — within those seven modules, two bars and two spaces of various widths make up the pattern that identifies the digit. There is a right-side bar-code pattern and a left-side pattern for each digit. That allows checkers to scan bar codes in either direction.

LEFT HALF	DIGIT	RIGHT HALF
0001101	0	1110010
0011001	1	1100110
0010011	2	1101100
0111101	3	1000010
0100011	4	1011100
0110001	5	1001110
0101111	6	1010000
0111011	7	1000100
0110111	8	1001000
0001011	9	1110100

0 1 2345 67890 (5)

1 2 3 4 5 6 7 8 9 10 11

POSITIONS

CHECK DIGIT

Checking Accuracy

The check digit (the last digit on the right) is used to make sure that a U.P.C. has been scanned correctly. The computer does the following calculations; if it comes out with the check digit, the scanning was correct.

For a U.P.C. that starts with 0 12345 67890:
Add the digits in odd positions: 0 + 2 + 4 + 6 + 8 + 0 = 20
Multiply the results by 3: 20 X 3 = 60
Add the digits in even positions: 1 + 3 + 5 + 7 + 9 = 25
Add the last two results: 60 + 25 = 85
Subtracting that answer from the next-highest multiple of 10 should produce the check digit: 90 - 85 = 5

Sources: NCR; the Uniform Code Council Inc.

Mika Gröndahl

Bar Codes: Reading the Patterns of Commerce

By Catherine Greenman

What distinguishes a 20-ounce bottle of Coke from a 2-liter bottle at the cash register? How can manufacturers and retailers understand what products sell better in different parts of the country? For the answer, look to the ubiquitous black and white stripes of bar codes.

A bar code label uses parallel black and white lines to encode information about a product. In stores, bar codes are read by a scanner, which translates the light reflected by their white spaces into electrical signals. Those are converted into digital signals and transmitted to the store's computer system.

Different bar-code languages, called symbologies, are used to identify different types of products, like groceries, clothing and electronics. The central group for bar code symbologies is the Uniform Code Council, an independent organization in Dayton, Ohio, that assigns and regulates all bar codes used in this country. The council cooperates with similar groups in Europe and Japan.

Manufacturers work with the council voluntarily, but most retailers have come to rely very heavily on bar codes — so much so, in fact, that a cashier has a hard time selling you something that doesn't have a bar code on it.

Although many bar-code languages have been developed by the retail and manufacturing industries since they were first adopted in the early 1970's, the most widely used type is the Universal Product Code, found in most groceries and retail stores. The U.P.C. is a 12-digit code: the first digit identifies the general category of the product; the next five, the product's manufacturer; the next five, the individual item (like Coke, Diet Coke or Caffeine Free Coke); and finally a ''check digit'' that is used to make sure that the code is scanned correctly.

The U.P.C. allows the store's computer to pull up other in-formation about the product, like price, inventory count and taxes. Because the U.P.C. links to the store's database instead of encoding the price itself, the price can be changed without fiddling with price tags for each individual item with a bar code.

There are a few other bar-code languages, although none are as prevalent as the U.P.C. Code 128 and Code 39, for example, were designed for printing on corrugated material like boxes. Code 128 and Code 39 are alphanumeric, so they represent letters as well as numbers.

All those codes are one-dimensional, or linear, meaning the information is read along a horizontal axis. But some bar codes are two-dimensional: they have vertical information as well as horizontal. There are two types of two-dimensional codes, stacked and matrix. In a stacked code, bar codes are arranged in vertical or horizontal rows. A matrix code uses vertical and horizontal squares and hexagons to embed the data.

The cost of printing and reading two-dimensional codes is too high for retailers, who need only basic product information. But two-dimensional bar codes are used in the transportation and package industries, which need to get access to databases in remote locations. Because two-dimensional codes can stack, or layer, up to 2,000 characters in a very small space, they can function as their own traveling databases.

Many states, including New York and Georgia, print a two-dimensional matrix code, developed by Symbol Technologies, on the back of driver's licenses. By scanning that code, a traffic officer can automatically enter all the information on the front of the license into a database instead of copying it by hand.

Supermarket Sweep

The typical checkout scanner uses a series of fixed and rotating mirrors to sweep a laser beam across the bar code on a package. The pattern of reflected light is detected by a sensor and turned into a digital code.

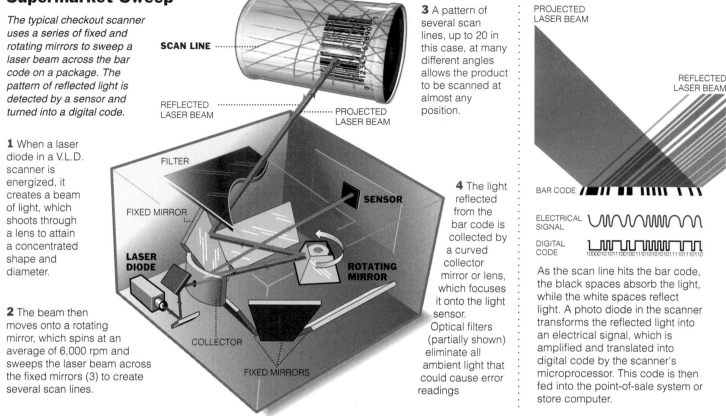

1 When a laser diode in a V.L.D. scanner is energized, it creates a beam of light, which shoots through a lens to attain a concentrated shape and diameter.

2 The beam then moves onto a rotating mirror, which spins at an average of 6,000 rpm and sweeps the laser beam across the fixed mirrors (3) to create several scan lines.

3 A pattern of several scan lines, up to 20 in this case, at many different angles allows the product to be scanned at almost any position.

4 The light reflected from the bar code is collected by a curved collector mirror or lens, which focuses it onto the light sensor. Optical filters (partially shown) eliminate all ambient light that could cause error readings

SCAN LINE
REFLECTED LASER BEAM
PROJECTED LASER BEAM
FILTER
FIXED MIRROR
SENSOR
LASER DIODE
ROTATING MIRROR
COLLECTOR
FIXED MIRRORS

PROJECTED LASER BEAM
REFLECTED LASER BEAM
BAR CODE
ELECTRICAL SIGNAL
DIGITAL CODE

As the scan line hits the bar code, the black spaces absorb the light, while the white spaces reflect light. A photo diode in the scanner transforms the reflected light into an electrical signal, which is amplified and translated into digital code by the scanner's microprocessor. This code is then fed into the point-of-sale system or store computer.

Sources: Metrologic Instruments, Inc.; Personal Information & Processing Services, Inc.; NCR Corporation

There's More Than One Way to Scan a Bar Code

By Catherine Greenman

In the bad old days, it was inevitable that someone in front of you in the grocery checkout line would hold things up by presenting an impossible-to-scan item to the cashier. No matter how many times it was swiped past the scanner, that jar of peanut butter just wouldn't go through.

But scanning technology is better now; virtually all bar codes are read successfully on the first pass. So successfully, in fact, that you barely have time to read The National Enquirer headlines as you move toward the register.

All scanners read a bar code in essentially the same way: by sweeping a beam of light across it and translating the pattern of black and white spaces into a binary code of zeros and ones that the store's computer can recognize.

Pen scanners, also called wand scanners, developed in the early 1970's, were the first to be widely adopted by retailers, and today they are still the least expensive. Pen scan-

ners use a laser-emitting diode light source at the tip of the pen and a light detector in the barrel, and they need to touch the entire bar code to read it. This is not a problem for scanning a box of detergent, but it is potentially cumbersome for scanning a can of soup.

Laser scanners, on the other hand, don't need any contact with the product and are able to read a bar code up to several feet away. They can be either handheld or stationary units, like the ones built into grocery store checkout counters.

Laser scanners built in the 70's and 80's used large glass tubes filled with helium and neon gas to send a laser beam across a bar code. Although these helium-neon laser scanners sped up the checkout process somewhat, they were heavy, generated a great deal of heat and had short life spans because the tubes broke frequently.

Visible laser diode, or V.L.D., scanners, which were intro-

More Than Just the Price

When a checkout scanner reads a bar code on an item, it displays the price to the customer and adds it to the bill. But the bar code data is also used by the store, and by other links in the retail, wholesale and manufacturing chain.

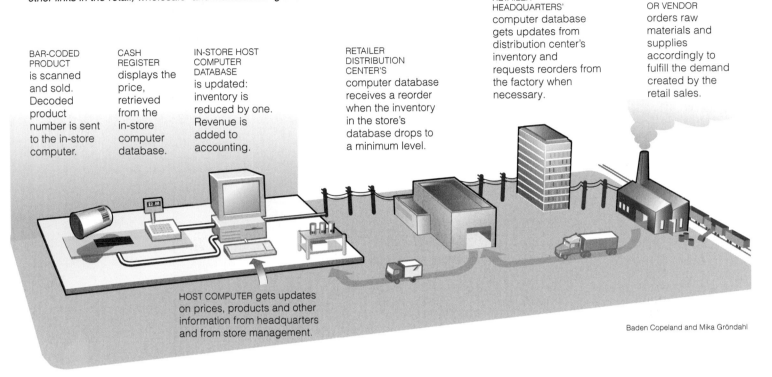

BAR-CODED PRODUCT is scanned and sold. Decoded product number is sent to the in-store computer.

CASH REGISTER displays the price, retrieved from the in-store computer database.

IN-STORE HOST COMPUTER DATABASE is updated: inventory is reduced by one. Revenue is added to accounting.

RETAILER DISTRIBUTION CENTER'S computer database receives a reorder when the inventory in the store's database drops to a minimum level.

RETAILER HEADQUARTERS' computer database gets updates from distribution center's inventory and requests reorders from the factory when necessary.

MANUFACTURER OR VENDOR orders raw materials and supplies accordingly to fulfill the demand created by the retail sales.

HOST COMPUTER gets updates on prices, products and other information from headquarters and from store management.

Baden Copeland and Mika Gröndahl

---

duced in the late 80's and are used in most supermarkets today, use a small diode to emit a laser beam. Because these diodes generate less energy than the helium-neon tubes and are a fraction of the size, V.L.D. scanners are able to emit anywhere from 15 to 50 different laser lines simultaneously, at several different angles, which gives the laser repeated opportunities to read the bar code, cutting down on unreadable items. The light beam produced by these scanners is also highly concentrated and will not disperse after it crosses a given distance. This means that an item does not have to line up perfectly as it is swiped or dragged across the scanner.

The third type of scanner, which uses a charged coupled device, is not swiped across an item but instead is pointed in front of it as an array of bright laser-emitting diodes collect the bar code's reflected light and create a video image of it. The signals from the image are decoded in the same way as with other scanners, by translating the image of the light pattern into a binary code of zeros and ones. Because it reads paper tags well, the charged-couple device is in common use in clothing and department stores.

Once the scanner reads and decodes the bar code, it

sets in motion the pricing and inventory process. The code identifies the product's manufacturer and individual type, and points the retailer to a specific record in the store's database. The current price of the item is pulled up from this database and returned to the cashier, where it is printed on the customer's receipt.

Hundreds of companies, including I.B.M. and Computer Associates, make software that lets retailers build their pricing and inventory databases. With these databases, when an item is scanned, the inventory level at that store is automatically reduced by one unit. The store computer is often pre-programmed to generate a reorder when the inventory count goes below a specified level.

Outside of the retail industry, laser scanners, which can shoot beams over a relatively large distance, are positioned over conveyor belts to keep inventory on individual product parts for manufacturers. Large mailers like Federal Express and the United Parcel Service also use overhead laser scanners pointed at high-speed conveyor belts to record package information.

Moving Pictures, No Moving Parts

The motion-image card — one type of lenticular image — is a familiar novelty. The cards may show a batter swinging, a dancer dancing or a smile changing into a frown. The image seems to move because the eye is tricked into seeing different images as the angle of sight changes. At least six images — and often dozens — are put on a card, but for the sake of simplicity only two are shown here.

CARD DECONSTRUCTED

If you draw a fingernail across the surface of a motion image card, you'll feel ridges in the plastic. These are actually rows of optical grade lenses that focus light on different strips of the surface at the back of the card and magnify whatever is there. When you look at the card from different angles – by moving the card or your head — you see different images, so the figure on the card changes or seems to move.

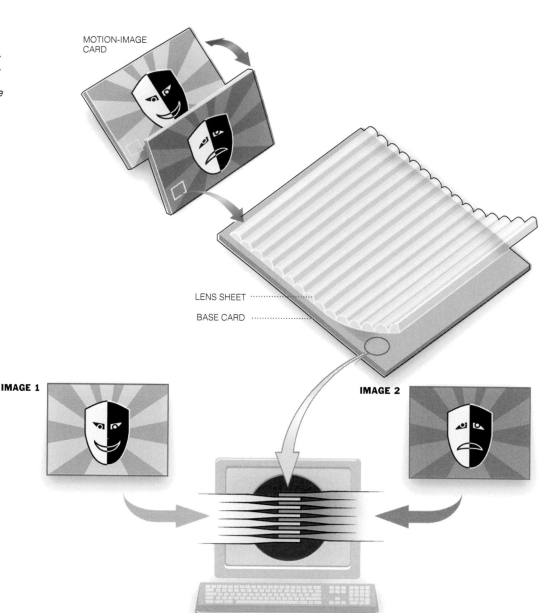

MOTION-IMAGE CARD

LENS SHEET

BASE CARD

IMAGE 1

IMAGE 2

SEEING THE MOTION

As your eyes move across a lenticular image, looking at it from different angles, or as you tilt the card, you see motion because each lens focuses on a different strip of image beneath it and magnifies it when the angle changes. To the eye, the strips merge to look like a whole image, although poor printing or poor alignment with the lenses can produce several shadowy images at once.

LENS

BASE

STRIP OF IMAGE 1

STRIP OF IMAGE 2

INTERLEAVING THE IMAGES

To make a lenticular image, as motion-image and autostereo cards are called, each image is divided into parallel strips (optically, by photographing it through a lenticular sheet, or mathematically, using software) and distorted so each strip is squashed thinner. The images are interleaved so one very thin strip of each image is printed next to a strip from the next. That requires very precise positioning, especially since most motion-image prints are color photographs that need to be printed in four separate color passes. The more sophisticated lenticular images can have dozens of images aligned beneath each lens. To make sure that these precisely printed images are precisely aligned with each lens, they are often printed in mirror image directly onto the back of the lens sheet.

Source: Depthography

Mika Gröndahl

Lenticular Images: It All Depends on Your Point of View

By Matt Lake

You may not have heard the term before, but the chances are strong that you've seen a lenticular image or two. In fact, its almost impossible not to go past one without stopping, cocking your head and swaying back and forth to get a better look. They've been stuck on cereal boxes, advertising awnings, department store displays, postcards and buttons and even, in a very crude form, on CD covers. And in every case, they contain some kind of eye-catching animation or a 3-D perspective on a picture, or both.

Since these images don't need a special viewer to display their optical illusions, they are sometimes called autoanimated and autostereo images. Some go so far as to call animated 3-D images 4-D images because they add the dimension of time.

The most common misperception is that these pictures are holograms. They're not. While both types of technology can add depth to two-dimensional pictures and create animated effects, they look different, work differently and are used for different purposes. Holographic images are created by lasers, can be printed on very thin metallic or transparent sheets and typically display a rainbow sheen like the reflection on a puddle of oil.

Lenticular images are much thicker because they have a ribbed plastic surface, and they are generally clearer and brighter than holograms.

Lenticular images have been around since just after World War II, when developments in plastics made it possible to create the ribbed sheet that sits on top of every motion-image card and autostereo image. These ribbed sheets are called lenses, or lenticular sheets, because they are actually dozens of optical-grade cylindrical lenses sitting in parallel lines on top of several specially prepared images. The lenses refract the images beneath them so that as your point of view changes, the image you see changes — creating the illusion of movement.

The three-dimensional effect works on the same principle as Viewmaster stereoscopes: Each eye sees its own version of the same scene, each taken from a slightly different angle, so the brain perceives the scene in three dimensions. On a lenticular sheet, each cylindrical lens refracts the different pictures to different eyes without the need for the viewer to look into a device.

From the outset, the lenticular technique was used for gee-whiz value. The company that pioneered the process in the 1940's, Variview, started by making animated political campaign badges with the slogan "I Like Ike!" and moved on to animated cards that were stuck on boxes of Cheerios. The technique was apparently too successful: cards were so frequently ripped off the boxes that the company had to put them inside the boxes.

Throughout the 1950's and 1960's, the lenticular process was used to produce sublime and ridiculous images, from flicker-image novelty rings and animated postcards of a winking face to a 3-D devotional picture of Jesus. In recent years, since Variview closed its doors, other commercial studios have continued the tradition.

Precisely how lenticular images are prepared is a closely guarded secret. Variview did not patent the process because its founder, Victor Anderson, didn't want to reveal his trade secrets. The original lenticular cameras expose images to photographic film through the numerous lenses of a lenticular sheet, which creates an interleaved image that is ready to develop photographically and apply to lenticular sheets. Now a lot of lenticular images start life as three-dimensional computer models or flat digital photographs, which are processed by computer and printed on commercial printers.

Not all lenticular images are created equal. The earliest examples were literally autostereo images: two images that either gave a parallax perspective on the same image or switched between two images.

But advertising-quality lenticular images cram dozens of frames into a single image. Beneath each lens, for example, could be lines from 30 to 50 images. The lenses magnify each line; together, they form a full-size image. If the sight of a lenticular picture isn't enough to make you stop in wonder, the fact that it's made up of that many different images surely should be.

Tiny Capsules All in a Row

The system developed by E Ink consists of microcapsules filled with a dark, inky liquid. Also inside the microcapsules, which are about a tenth of a millimeter in diameter, are even smaller particles — white beads. The microcapsules are sandwiched between an opaque bottom layer and a transparent top layer. Energized by an electric charge, they form patterns that make up letters, like those in the message at right.

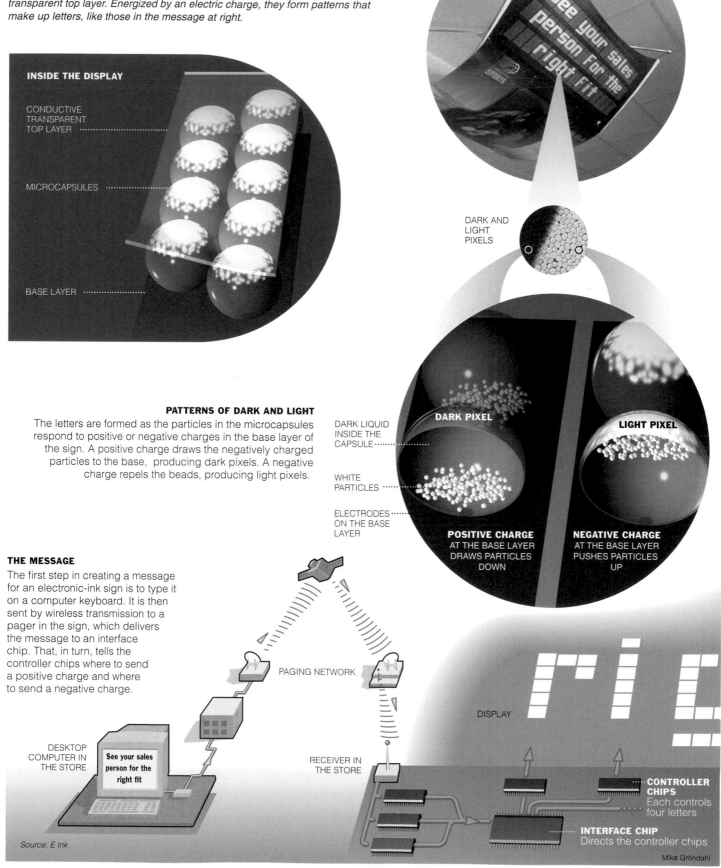

INSIDE THE DISPLAY

CONDUCTIVE
TRANSPARENT
TOP LAYER

MICROCAPSULES

BASE LAYER

DARK AND
LIGHT
PIXELS

PATTERNS OF DARK AND LIGHT

The letters are formed as the particles in the microcapsules respond to positive or negative charges in the base layer of the sign. A positive charge draws the negatively charged particles to the base, producing dark pixels. A negative charge repels the beads, producing light pixels.

DARK LIQUID
INSIDE THE
CAPSULE

WHITE
PARTICLES

ELECTRODES
ON THE BASE
LAYER

DARK PIXEL

LIGHT PIXEL

POSITIVE CHARGE
AT THE BASE LAYER
DRAWS PARTICLES
DOWN

NEGATIVE CHARGE
AT THE BASE LAYER
PUSHES PARTICLES
UP

THE MESSAGE

The first step in creating a message for an electronic-ink sign is to type it on a computer keyboard. It is then sent by wireless transmission to a pager in the sign, which delivers the message to an interface chip. That, in turn, tells the controller chips where to send a positive charge and where to send a negative charge.

PAGING NETWORK

DESKTOP
COMPUTER IN
THE STORE

See your sales
person for the
right fit

RECEIVER IN
THE STORE

DISPLAY

**CONTROLLER
CHIPS**
Each controls
four letters

INTERFACE CHIP
Directs the controller chips

Source: E Ink

Mika Gröndahl

Electronic Ink: A Changing Medium Is the Message

By Lisa Guernsey

Gutenberg would turn over in his grave. Not at the notion of electronic ink, which can rearrange itself in response to electrical charges, allowing a page of text to change from "Hamlet" to "Windows for Dummies" at a programmer's whim. No, he would turn over at some of the first words in the new ink to reach the public.

They hung on a sign in the sportswear department of a J.C. Penney store in Marlborough, Mass. They read: "See your salesperson for the right fit."

Not exactly poetry, is it? But a commercial message on a large sign was a perfect test case for the technology, according to the founders of E Ink, the company that created the J.C. Penney signs. The signs were large enough to allow the developers to experiment without having to tinker in tiny spaces with tiny materials. And a market for the technology already existed: store managers wanted light, flexible signs that could be programmed to flash catchy new messages without sucking up too much electricity.

The J.C. Penney sign was the beginning, said Paul Drzaic, director of display technology for E Ink, which is based in Cambridge, Mass. "The end point is the newspaper or book."

E Ink was founded by four men: Joseph Jacobson, a professor at the Media Laboratory at the Massachusetts Institute of Technology; two of his students, Barrett Comiskey and J.D. Albert; and Russell J. Wilcox, a graduate of Harvard University's business school, who joined the others after writing his thesis on how to build a company around electronic-ink technology. Dr. Jacobson conducts research on electronic displays — and much of his research forms the basis of E Ink's signs.

Across the continent, Nicholas Sheridon, a researcher at Xerox's Palo Alto Research Center, has experimented with similar technology, creating what he calls electronic paper. Both it and electronic ink use the same principle: ink reacts to an electric charge. A positive charge causes one color to appear. A negative charge causes a different color to appear. Letters are spelled and graphics are displayed by the pattern of those two colors.

In E Ink's design, the ink is composed of white particles suspended in an oily blue dye inside clear microcapsules, each a tenth of a millimeter in diameter – small enough to fit inside the period at the end of this sentence. Each white particle within the microcapsule carries a slightly negative charge. To the human eye, the ink looks like a smooth liquid.

The microcapsules are sandwiched between a base that is coated with graphite and has electrodes connected by tiny silver leads to computer chips, and a thin, transparent layer of indium tin oxide, which is the side turned toward the viewer.

The computer chips provide the instructions to produce negative and positive charges across the display, depending on the ink patterns that are desired. When a section of the display becomes positively charged, for example, the tiny white particles inside the microcapsules are drawn to the base, leaving the blue dye on top, producing dark areas or pixels in the sign. In other sections the white particles might be forced toward the top, creating light pixels that form the letters on the dark background.

Only a short application of voltage is needed to rearrange the ink, so the signs use less power than video screens or computer monitors, which require a steady stream of electricity to display a message.

Dr. Sheridon, the researcher at the Xerox center, is working on a similar problem – but instead of testing his technology on commercial signs, he is focusing on the creation of electronic paper. He has been developing thin sheets of tough rubber that can be manufactured to be almost as thin as paper. Thousands of tiny plastic beads are stuck in the rubber sheet, suspended in an oily ink that is also encased within the sheet. The beads have two colors: one hemisphere is white, the other black.

Two conductive plates are applied to the back and front of the rubber sheets, enabling the conduction of electric charges. A positive charge will cause the white side of the beads to rotate upward; a negative charge will bring up the black side. As with E Ink's signs, the patterns of white and black can then create letters that form words and sentences.

On the Disassembly Line

In a typical recycling plant, the machines that clean, sort and grind the bottles are arranged in a line for efficiency. These systems can handle thousands of bottles each hour, which can create storage headaches. "People ask me, 'How big is your machine?'" says Charles Roos, board chairman of National Recovery Technologies, which makes sorting equipment. "I say, 'How big is the warehouse in front of it and the warehouse behind it?'"

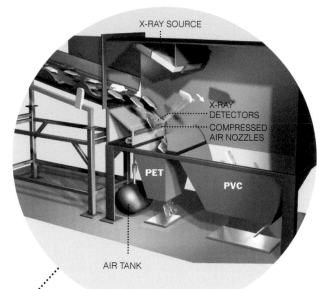

X-RAY SOURCE

X-RAY DETECTORS
COMPRESSED AIR NOZZLES

PET

PVC

AIR TANK

BREAKING BALES
Bales of compacted plastic bottles from municipal collections or bottle-return centers are broken up.

SCREENING
Dirt and loose labels are removed.

REMOVING PVC PLASTIC
X-rays and compressed air are used to separate bottles made from polyvinyl chloride, or PVC, from those made from polyethylene terephthalate, known as PET or PETE. PVC absorbs more X-rays than PET; the difference is detected by a sensor linked to a computer. When a PVC bottle passes, the computer activates a solenoid valve, and a blast of compressed air blows the bottle from the stream.

WASHING
Bottle washing is common in Europe, but not in the United States.

SORTING BY COLOR
Visible light is used to separate PET bottles by color. The machine can also separate PET bottles from those made from HDPE (high-density polyethylene).

GRINDING
Grinds the bottles into small flakes, about three-eighths of an inch in diameter.

SORTING FLAKES
Infrared detectors are used to identify remaining PVC and other plastics in a stream of PET flakes, and compressed air removes the contaminants. The PET is then ready to be made into something else.

Recycling symbols: a bottom-of-the-bottle guide to plastics

PETE
Polyethylene terephthalate
Soft-drink bottles

HDPE
High-density polyethylene
Laundry detergent bottles, grocery bags

V
Polyvinyl chloride
Shampoo bottles

LDPE
Low-density polyethylene
Bread bags

PP
Polypropylene
Margarine tubs, ketchup bottles

PS
Polystyrene
Meat trays, egg cartons

OTHER
Polycarbonate, acrylic, ABS, others

Sources: National Recovery Technologies Inc.; American Plastics Council

Frank O'Connell and Mika Gröndahl

Recycling Sorters Ensure That Garbage In Isn't Garbage Out

By Catherine Greenman

In the curbside afterlife, the road from plastic Coke bottle to cozy polar fleece is paved with dirt, caps, sticky labels and extraneous resins, all of which have to be removed before a bottle can be recycled.

To make that recycling possible, increasingly sophisticated technology helps bottle recyclers wash, sort, grind and analyze the bales of crushed bottles they buy from waste haulers or municipalities.

The makers of recycling equipment include National Recovery Technologies and MSS Inc., both based in Nashville, and SRC Vision, in Medford, Ore. The companies have developed systems to detect and separate different plastic polymers using X-ray, infrared, near-infrared and visible-light sensors.

The systems allow recyclers to process thousands of bottles an hour and to produce very precise separations of the different kinds of plastic resins used in bottle manufacturing. To be profitable, plastic recyclers need to produce clean, reusable batches of a single type of resin. The resins are used to make products like polyester carpet, insulation for winter coats and new bottles.

The production of commercially viable polyethylene terephthalate, known as PET or PETE, is the mainstay for many bottle recyclers. PET is used in most plastic soda bottles; it is also used to package products like peanut butter, salad dressings and household cleaners.

The recycling of PET bottles is greatly complicated when a batch is contaminated by bottles made with polyvinyl chloride, or PVC. PVC bottles, which are typically used for household cleaners and personal care products like shampoos, currently account for 2 percent of the market. Even small traces of PVC in a batch of PET, or traces of PET in a batch of PVC, will cause structural problems in recycling.

Although some recycling companies attempt to sell their PVC stock, it has less resale value than PET, which is currently worth about 12 to 13 cents per pound. "We've tried to sell it in the past for a couple of cents a pound," said Doug Meredith, a project engineer at Mohawk Industries, a carpet manufacturer in Calhoun, Ga., speaking of PVC. "There's really not much of a market for it, and it usually goes into a landfill." The company uses PET in carpets.

When PVC is not recycled, it winds up in landfills or is incinerated, which leads to the formation of dioxin, an environmental hazard. Environmental groups like Greenpeace maintain that it would be better to stop producing PVC instead of putting resources into removing it with sophisticated and expensive PVC removal systems.

But Allen Blakey, a spokesman for the Vinyl Institute in Arlington, Va., said bottle manufacturers used PVC because its flexibility made it easier to mold handles. Blakey added that if PVC was incinerated at high temperatures, the release of dioxin could be controlled.

In the early 1990's, it was common to see employees at a recycling plant separating bottles by color, or looking on the bottom of each one for a triangular symbol signifying the type of plastic.

Now automated bottle separators can separate bottles according to their plastic content or color, so they have augmented or replaced human sorters. National Recovery Technologies and MSS manufacture systems that separate bottles by color by measuring how clear, blue and green bottles respond to light.

And the VinylCycle system of National Recovery Technologies, for example, aims X-rays at bottles on a conveyor belt to detect the chlorine atoms that indicate PVC bottles. The VinylCycle's computer then triggers a puff of air that separates the PVC bottles from the others.

Some recycling companies buy bottles that have been separated by color and ground into small flakes that are primarily PET. Gary Pratt, the president of one of the largest mixed-plastics recyclers in the country — P & R Environmental Industries, based in Youngsville, N.C. — said it was important to purify the PET flakes. "PET is the dominant resin we're looking for," he said, "but if the PVC levels are greater than 100 parts per million in our PET flake, the load could be rejected by the company we sell it to."

The FlakeSort system, also made by National Recovery Technologies, funnels plastic flakes along a conveyor toward an infrared-based detection system that looks for the light absorption characteristics of the different plastic polymers. If a plastic flake is not PET, air nozzles send it flying, like a bad egg, away from the rest.

A Place for Everything, and Everything in Its Place

Radio waves used for communications and research lie in part of the electromagnetic spectrum, which also includes infrared, ultraviolet and visible light, X-rays, gamma rays and cosmic rays. Radio waves have frequencies ranging from about 3 kilohertz, or 3,000 cycles per second, to about 300 gigahertz, or 300 billion cycles a second.

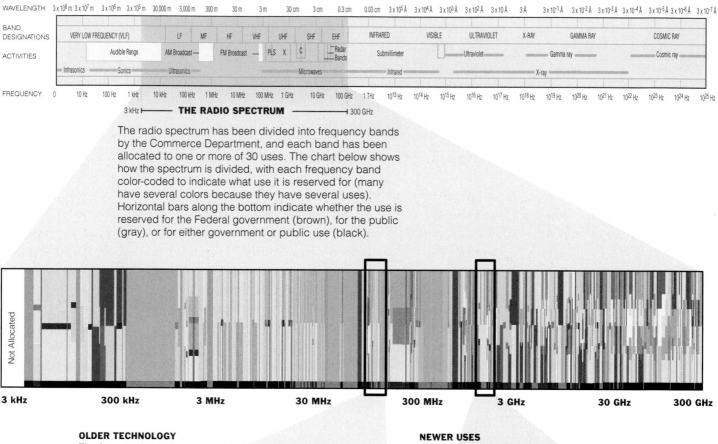

THE RADIO SPECTRUM — 3 kHz to 300 GHz

The radio spectrum has been divided into frequency bands by the Commerce Department, and each band has been allocated to one or more of 30 uses. The chart below shows how the spectrum is divided, with each frequency band color-coded to indicate what use it is reserved for (many have several colors because they have several uses). Horizontal bars along the bottom indicate whether the use is reserved for the Federal government (brown), for the public (gray), or for either government or public use (black).

Not Allocated

3 kHz · 300 kHz · 3 MHz · 30 MHz · 300 MHz · 3 GHz · 30 GHz · 300 GHz

OLDER TECHNOLOGY

The frequencies at the lower end of the spectrum have been in use the longest. Among the frequencies shown below are the ones used for communication between airplanes and air traffic controllers (aeronautical mobile); for satellite communications and space research; for two-way radios (mobile and fixed); for amateur, or "ham," radio; for ship-to-shore communications (maritime mobile); and for broadcast television Channels 7 through 13.

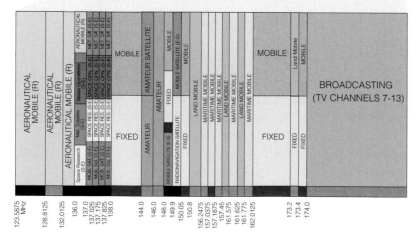

NEWER USES

Technological advances have opened up the higher end of the radio spectrum for new uses. Among them are: transmitting images and data from weather satellites (shown here as "met sat," short for meteorological satellite), transmitting signals from PCS (personal communications services) devices (the widest "fixed" band below), broadcasting digital radio signals (broadcasting satellite), and transmitting signals from Global Positioning System satellites (radiolocation).

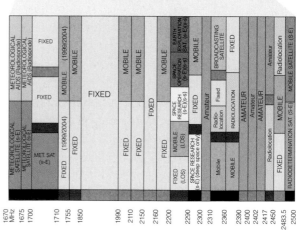

Source: U.S. Department of Commerce, National Telecommunications and Information Administration, Office of Spectrum Management

Mika Gröndahl and Juan Velasco

Directing Traffic in the Radio Spectrum's Crowded Neighborhood

By Dylan Loeb McClain

It is a growing, although invisible problem.

Since the early 1990's, an explosion in the number of wireless devices — cellular telephones, satellite television dishes and pagers, to name just a few — has led to crowded airwaves.

So far, the range of radio frequencies has been able to accommodate all the increased uses. More and more, though, ensuring that different devices can be used without interference from others is a high-wire act achieved only through technological advances and the careful management of the radio spectrum.

Different uses are allotted different portions of the usable spectrum, from relatively low frequencies (measured in thousands of hertz, or kilohertz) to high-frequency microwaves (measured in billions of hertz, or gigahertz). The result is a crazy quilt of frequency bands reserved for a variety of civilian and military applications, like conventional AM and FM radio, television, ham radio, mobile phones, aircraft and ship communications, and navigation systems. Some frequency bands are left unused to avoid interfering with radio astronomy and other scientific efforts.

The International Telecommunication Union, a United Nations organization that is the intergovernmental rule-making body for the radio spectrum, used to review how radio frequencies were used every 20 years. Since 1992, though, it has stepped up its reviews to every two years to keep up with the demand.

In the United States, the agencies responsible for managing the spectrum have never been busier. They are the National Telecommunications and Information Administration, which handles the parts of the spectrum reserved for the government's use, and the Federal Communications Commission, which keeps watch over private use.

"As you make more of the spectrum available, more people want to use it," said Norbert Schroeder, a program manager in the office of spectrum management at the telecommunications agency. Decisions about how to use the spectrum hinge on practical questions — what it is possible to do — as well as on public policy, international regulations and economics, Mr. Schroeder said. Technology has come to play a crucial role. "It is only because of the developing technologies that the spectrum has become available at the higher frequencies," Mr. Schroeder explained.

But tapping into ever-higher frequencies has its own problems. The signals at higher frequencies are more easily obstructed by buildings and other structures. To use these frequencies, which start around three gigahertz and are mostly in the microwave part of the spectrum, it is necessary to use relays to strengthen and retransmit the signals or to build very big antennas. Both solutions are expensive, Mr. Schroeder said.

Dale Hatfield, director of the Federal Communications Commission's office of engineering and technology, said an alternative approach was to use the lower frequencies more efficiently.

Here, too, technology can help. Digital systems make it possible to compress signals, reducing the width of the frequency bands needed for some transmissions. For example, the frequency bands allocated for two-way radios for taxis and the police used to be 50 kilohertz wide. Now they are just a fraction of that. Still, there are trade-offs. "With the smaller bands, you lose transmission power," Mr. Schroeder said. "But system design can overcome these problems by using relays or repeaters that rebroadcast the signals."

Sharing frequencies is another way to manage the demand. The radio broadcasts of the Voice of America, for example, share frequency bands that are used by the military for training.

Reallocating what is already in use is a possibility. The part of the spectrum used for U.H.F. (ultrahigh frequency) television channels 60 to 69 ranges from 746 to 806 megahertz. That band is not in use in some parts of the country, so part of that chunk of the spectrum in those areas was sold at auction to companies that provide wireless Internet communications for personal digital assistants, cell phones and personal computers.

Even with changes and technological developments, spectrum management remains difficult. One strain on the system is unlicensed use. The spectrum from 902 to 928 megahertz, for example, is occupied by signals that do not need to be licensed, like those coming from cordless phones and keyless security systems for doors.

"People a lot of times are using radio spectrum without realizing it," Mr. Hatfield said. And equipment that makes no use of radio frequencies can also cause interference. "The spectrum is a natural resource that can be polluted," Mr. Hatfield said. "A person turning on a hair dryer can cause snow on the TV next door."

Transportation

A Salad Spinner for Cars

The National Advanced Driving Simulator, located at the University of Iowa, is designed to emulate all kinds of driving experiences realistically. The centerpiece is a computer-controlled projection dome containing a modified vehicle that can pitch, wobble and spin on its own, and can travel on a track system for realistic driving simulations.

PROJECTORS
Project realistic front and rear-view images on screens that wrap around the inside of the dome. ······

ACTUATORS
Actuators replace the wheels on test vehicles, connecting to the axles to provide bumps, shakes and a realistic feel when turning.

TURNTABLE
Allows the entire dome to spin, increasing realism during spins and other accident simulations.

HYDRAULIC LEGS
Similar to those found in airplane simulators, the legs contain pistons that move rapidly up and down, simulating uphill or downhill travel, rough terrain or accidents.

Simulating a spin

The dome is propelled by a moving belt and rides on two rails, allowing it to travel back and forth along one axis. This belt-rail system can travel back and forth along a perpendicular axis on another system of rails and belts. The net effect is that the dome can travel in any direction across the floor, spinning, dipping and wobbling at the same time.

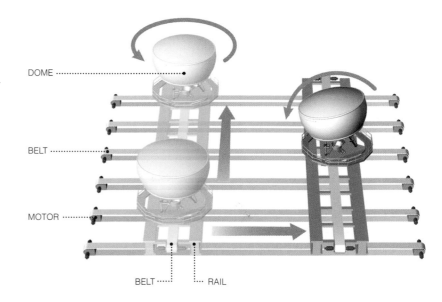

DOME ········

BELT ········

MOTOR ········

BELT ······ ···· RAIL

THE TESTING ROOM

The dome is contained in a room about the size of a high school gymnasium. On one wall is a glass-enclosed control booth. To change test vehicles or subjects, the dome is maneuvered up to access doors in the wall.

Sources: University of Iowa; National Highway Traffic Safety Administration

Frank O'Connell

Driving Simulator: Ersatz Chills and Spills in the Name of Safety

By Melinda Pradarelli

A new vehicle that rolled off the assembly line in 2000 has turned some heads, despite the fact that it will never be street-legal. Weighing in at three tons and lacking a gas tank, it nevertheless leaves scientists, engineers and lawmakers across the country drooling.

It's the National Advanced Driving Simulator, a project of the National Highway Traffic Safety Administration and the University of Iowa, and it began operation in Iowa City after eight years of planning and construction.

The simulator, called NADS, is a research tool to improve vehicle and highway design and better predict driving behavior — all with the goal of saving lives. It gives researchers the chance to test car safety systems, for example, or ride virtual shotgun with drivers whose abilities are impaired by alcohol or medication or who are so absorbed in a cell-phone conversation that they do not notice that they are driving off the road.

"There has never been a motion simulator as complex or as large as the NADS," said Keith Brewer, director of the Office of Human-Centered Research at the highway safety administration in Washington.

The simulator, which cost about $60 million, is the world's first 360-degree, high-fidelity sound and motion system. Its 24-foot-diameter projection dome — perched on six hydraulic legs above a track system — is designed to surround experimental subjects with realistic road images. The dome can house a variety of types of vehicles, ranging from cars and sport utility vehicles to trucks, buses, backhoes and even tanks.

Both businesses and government agencies — including the highway safety administration — will pay the university to perform tests with the simulator.

About one-third of the simulator's time is allotted to companies like car manufacturers, so they can design smarter cars, and to pharmaceutical companies, so they can test how new drugs will affect drivers. The remainder goes to the federal government to examine drivers' actions and reactions behind the wheel.

Data gathered by the safety administration since the 1970's show that 90 percent of highway deaths each year are caused by human errors, not mechanical malfunctions.

Mr. Brewer said poor decision-making, risk-taking, poor vi-sion, fatigue and other factors contributed to the accidents. But until the simulator was built, he said, researchers have had limited ways to quantify the causes. Historically, they either used a cruder simulator that did not allow for total immersion or conducted limited on-road tests with people in no traffic.

"NADS is different because it's so incredibly realistic," said Ginger Watson, the NADS branch chief of human factors and highway safety research at the University of Iowa. "We can expose people to a set of procedures when they are on a drug or off, with a device like a cell phone or without. Then we can put those same variables together over multiple drivers and record the results."

Researchers worked for four years to build a simulator that looks and feels like a real car. In fact, the module inside the dome can be transformed into a Ford Taurus, Chevy Malibu, Jeep Cherokee, Freightliner truck or dozens of other vehicles, using actual models that have been modified for testing purposes.

The hydraulics and track system, which take up a space larger than a basketball court, allow the simulator to move like a vehicle on a three-lane highway — skids, sharp turns, spins, lane changes and all.

The simulator accurately reproduces the visual and sensory cues of real-world driving. A computer system is used to control 15 projectors within the dome that can run dozens of different driving simulations. The computer system also records the driver's movements, taking 1,000 readings per second. It takes a snapshot of the vehicle's position and instruments 500 times a second.

There is also a second, more powerful computer that controls the simulator's motion base and directs everything in the car, like air conditioning, cell phone availability, low-grade engine noise, real traffic sounds and much more.

"There was an effort to change the limits last year," said Mr. Brewer, of the highway agency, "but they failed because there was a lot of lobbying by the alcohol industry, and we didn't have any rock-solid scientific evidence. This will change that." Mr. Brewer said he hoped that the new research would help the agency reduce the number of annual alcohol-related deaths to 11,000 by 2005.

Seeing the Night

Military night-vision technology made its way onto American roads as an option on Cadillac DeVilles. The system uses an infrared camera that detects heat emitted by people, animals or objects and converts it to a video image projected onto the windshield in front of the driver.

CONTROL SWITCH

Allows the driver to turn the system on or off, and change the intensity of the image with a slide lever. Another lever controls vertical position of the projection.

THE CAMERA

The Night Vision camera is mounted in the front of the engine compartment, where it can be safe from road debris. A lens system in the camera directs incoming infrared radiation onto a bed of barium-strontium-titanate capacitors, which store an electrical charge. The change in the charge for each detector is converted into a video signal, for display on the windshield.

HEADS-UP PROJECTOR

The video image, with hotter objects whiter in appearance, is projected onto the lower portion of the vehicle's windshield. The image has a field of view of 11 degrees horizontal by four degrees vertical, and is made to appear as if it is at the front edge of the vehicle's hood, thus avoiding the need to refocus the eyes to view it. To ensure the integrity of the image, only windshields free of optical distortion are installed in cars with the Night Vision system.

Due to legal restrictions, headlights may illuminate an area only about 150 feet in front of a vehicle. An infrared system allows objects to be seen up to five times that distance, as long as the objects emit heat. Even recently laid tire tracks in snow can be viewed, as they have some residual heat.

NIGHT VISION OFF

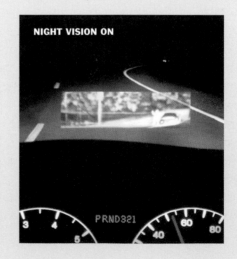

NIGHT VISION ON

John Papasian

Night Vision: Military Technology to Help Guide Your Personal Tank

By Eric A. Taub

One of Superman's most compelling talents was X-ray vision. The Man of Steel could see right through objects to the substance underneath. Now any of his fans who buy a Cadillac DeVille can experience something similar.

Customers can purchase a DeVille equipped with Night Vision, which allows drivers to see five times as far as they can usually see with headlights — and even lets them see objects hidden by darkness or obscured by trees. It works by projecting small images of heat-generating objects in a display across the lower part of the windshield, in front of the driver.

The system was developed by Raytheon for General Motors and is based on declassified military technology widely used in the Persian Gulf war. It is one of several approaches to improving drivers' vision at night without increasing the brightness or range of standard headlights.

"A driver's inability to see at night is independent of age," said Ed Zellner, chief engineer for the luxury car division at General Motors. "Twenty-eight percent of driving is done at night, but 55 percent of fatalities happen then." Raytheon approached Cadillac in the early 1990's with its night vision technology because it was looking for ways to expand its business beyond the military. The system was not close to being ready; it was cooled by liquid nitrogen and took up so much room it would have been too big for even the DeVille's gargantuan trunk.

While Raytheon worked on shrinking the system's size, Cadillac developed an easy-to-use interface. Engineers quickly decided not to put a display on the instrument panel because drivers would have had to take their eyes off the road to see what was ahead. Instead, they chose a virtual display that would make a deer, a pedestrian or a car in the road ahead appear to be in front of the car's hood. That eliminates the need to refocus one's eyes.

Perhaps most important, the image is the same size it would be and in the same relative position it would be in if you could see it unaided. Night scenes look like photographic negatives; the more heat an object gives off, the whiter it appears. Drivers can select the level of image brightness to accommodate their eyes' adaptation to darkness (think of what happens when you turn on a light inside the car).

The system uses a small infrared camera, mounted behind the car's grille. Infrared radiation – which is given off by humans, animals cars or any other object that produces heat — is absorbed by one of 76,800 detectors in the camera. Each detector element responds electrically based on how much infrared radiation it receives. The changes are processed to create a visual image, which is projected onto the display.

"The first reaction drivers have when they use Night Vision is 'Wow,' " Mr. Zellner said. "They can't believe the stuff that's out there in the darkness." The question for some is whether it's important to see everything that's there. "Sure, you can see deer, but they appear so quickly, an infrared system may not help you avoid them," said Jeff Erion, lighting manager for Ford's Visteon Automotive Systems, in Dearborn, Mich.

Mr. Erion says there is a cheaper way to help drivers see some objects beyond their headlamps: road signs could be coated with ultraviolet chemicals that would show up when illuminated by headlamp bulbs that emitted ultraviolet as well as regular light.

Visteon has also developed a "near infrared" system that enhances the perception of the area covered by normal headlights, rather than reaching far beyond that zone. An infrared transmitter sends out a beam that bounces off surrounding objects. A camera, similar to a video camera, detects infrared reflections from objects in the car's path and, through digital signal processing, turns them into an image that is projected directly across the entire windshield. That provides increased detail to the area already illuminated by headlights.

DaimlerChrysler has also taken the "active illumination" approach, building an infrared system that, like Visteon's, sends out an infrared beam, rather than just detecting infrared radiation from heat-emitting objects. "With our system, you can see objects, such as road markers and the road surface, that are the same temperature as the environment," said Peter Narozny, head of the company's high-frequency technology laboratory in Stuttgart, Germany.

Little Bulb, Big Light

Car headlight designs have evolved greatly since the mid-1980's, when federal standards governing them were eased. Newer headlights are brighter, use replaceable bulbs and are shaped to add to the aerodynamics of the car. Here's the current state of the art.

HALOGEN Like many bulbs, halogens use a glowing filament to produce light. First, an electric current passes through the tungsten filament, causing it to glow. Some tungsten evaporates, but halogen gas (bromine and iodine) helps return the tungsten to the filament. That reduces tungsten deposits on the inner surface of the bulb so the bulb stays brighter as it ages. Halogen headlights are much brighter than old-style headlights, but still only about half as bright as allowed by law.

HIGH-INTENSITY DISCHARGE H.I.D. lamps use tungsten-rod electrodes rather than filaments. A 10,000-volt charge is delivered to one electrode housed in a quartz glass capsule. A spark arcs through inert xenon gas to the other electrode and the lamp lights up. Then the current drops back to about 85 volts to keep the blue-white arc alight. H.I.D. lamps are much brighter than halogen and last about 10 times longer.

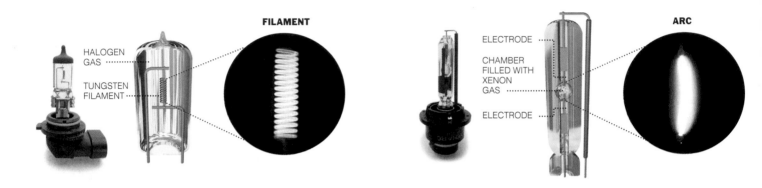

HALOGEN GAS

TUNGSTEN FILAMENT

FILAMENT

ELECTRODE

CHAMBER FILLED WITH XENON GAS

ELECTRODE

ARC

Creating the Beam

A headlight bulb sends light back toward an aluminum-coated reflector, which then reflects it toward the front of the lamp. Regardless of its source, the light has to illuminate the road evenly without blinding oncoming drivers. Here is how that can be done.

TRADITIONAL PARABOLIC HEADLIGHTS
A parabolic reflector collects light from the bulb and directs it to the lens, which can have up to 200 separate facets to create a desired beam pattern. Some light is lost in the thick molded glass or plastic lens.

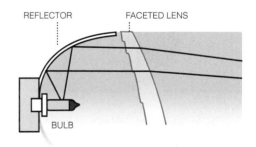

REFLECTOR   FACETED LENS

BULB

JEWELED HEADLIGHTS
Instead of having a light-shaping lens in the front of the headlight, the reflector itself is responsible for the correct beam pattern and direction. The reflector can have some 50,000 individually calculated points to direct the light from the bulb.

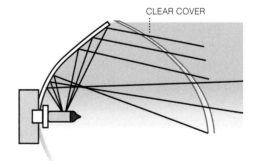

CLEAR COVER

PROJECTION-BEAM HEADLIGHTS
Light from an ellipsoid reflector is focused by a fish-eye lens, which projects it onto the road. The beam can be made smaller in diameter, making it look brighter. This type of light tends to illuminate the road near the car well but does not do as good a job with objects at a distance.

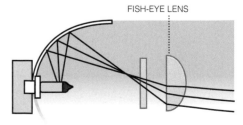

FISH-EYE LENS

Sources: Hella, Osram and Guide Corporations

Mika Gröndahl; Photographs by Tony Cenicola

For an Aging Population, Help With the Road Ahead

By Eric A. Taub

If driving at night has recently become more difficult, don't necessarily blame your headlights. There's a good chance that the problem has nothing to do with your car's equipment and everything to do with your age.

One sign of getting older is a loss of visual acuity. "When you're 45, you need 50 percent more light to see as well as you did when you were 25," said Jeff Erion, lighting manager for Visteon Automotive Systems in Dearborn, Mich.

Facing the largest group of geezers in history, the automobile industry has an economic incentive to make things brighter. The relaxation of federal regulations and the advent of powerful computer software have made it possible for the industry to create new headlight designs in record time.

In any headlight, the light needs to be properly dispersed. A headlight bulb generates light back toward an aluminum-coated reflector, which then reflects it toward the front of the headlight, which may have a plastic or glass lens. In a traditional parabolic headlight, up to 200 separate facets on the lens direct the light onto the road. The lens ensures that the beam is wide enough, evenly spread and aimed in the correct direction to illuminate road signs, while not shining in oncoming drivers' faces.

There are other designs as well: "jeweled" headlights do not use lenses, while projection-beam headlights do.

The newest trend is the jeweled, completely clear, headlight. Stare into a jeweled headlight (unlighted, if you're smart) and you can see through it to the back of the reflector. These lights "give the front of the car a more three-dimensional look," Mr. Erion said.

Until the 1980's, there was little designers could do to improve either a car headlight's output or looks because headlights, according to lighting engineers, were the most regulated part of a vehicle. Cars manufactured for sale in the United States had to meet stringent guidelines for headlight size, shape, placement and intensity. All lights had to be sealed beams, in which the lens, light and reflector are in one unit, and had to be mounted in either one or two pairs. A good result of such standardization was that it kept prices down for replacing headlight units.

By the mid-1980's, federal guidelines had been eased. Headlights can now be any shape – as long as the rules governing the lights' brightness, throw (how far the beam extends) and direction are followed. Ford's 1986 Taurus was one of the first models to use aerodynamic headlights that were flush with the sheet metal. And headlights no longer have to be replaced as a unit; lights can now use replaceable, brighter halogen bulbs, the kind that have been used by most manufacturers outside the United States.

The first aerodynamic headlights using replaceable bulbs had a different shape but performed no better than their sealed-beam predecessors. Engineers still needed to find an elegant compromise that would allow them to put enough light in the right places along the road without blinding oncoming drivers. Every time a new headlight shape is created, that process begins anew.

The lighting industry has toyed with a number of solutions to increase nighttime illumination. Manufacturers have experimented with headlights that can reflect ultraviolet light up to twice the distance of conventional lamps; but for such lights to be effective, all road signs (and pedestrians' clothing) would have to be coated with UV-reflective paint.

BMW, Lexus and other manufacturers have tried projection-beam headlights, which use fish-eye lenses to concentrate the light and make it appear brighter. While projection lamps are good at reducing stray light and glare from raindrops, "there's a great flood of foreground light with projection lenses, but not so much in the distance," Mr. Erion said.

The state of the art is high-intensity-discharge lighting. Rather than using a glowing filament to create light, H.I.D. lamps use high voltage to generate a bluish spark between two electrodes, exciting the xenon gas and leading to the emission of a bluish light. These headlights create brighter illumination within government standards. Because the light is blue, it appears brighter still. H.I.D. lights are still expensive; most are found on luxury cars.

Paying the Digital Way

The electronic payment of tolls keeps traffic moving along and spares drivers any frantic searches for dollars or dimes. And the dents it makes in drivers' wallets are hardly felt — until the statement of toll charges arrives in the mail each month. Here's how one electronic system, E-Z Pass, works.

ANTENNA AND READER

An automatic vehicle identification, or A.V.I., antenna sends out radio signals. The E-Z Pass tag in a car passing through the toll lane is responds to the A.V.I. signal with a signal that contains the tag's identification number. The A.V.I. reader, which can monitor up to eight lanes simultaneously, receives this identification information and relays it to the lane controller.

LANE CONTROLLER

The lane controller, a computer, searches its database to make sure that the identification number is for a valid account and that there is enough money in the account to pay the toll. If everything's all right, the car is waved through — a green light flashes or a barrier is raised. The lane controller also updates a computer at the toll plaza, which in turn updates the system's main customer service computer.

TRANSPONDER

The E-Z Pass tag, a small plastic box that is usually mounted on the windshield, is a transponder that can respond to radio signals. It contains a radio receiver, a radio transmitter, a dedicated logic circuit and a 10-year button battery. The transponder transmits a tag's identification number each time the car is driven through a toll lane.

VEHICLE CLASSIFICATION

The E-Z Pass system must figure out what kind of vehicle is in the lane so it knows how much to charge. The entry loop, which responds to magnetic changes, alerts the system to the vehicle's approach. The weight of the wheels rolling over the electromechanical treadle embedded in the road allows the system to determine the number of axles. And an overhead laser profiler uses information about how the vehicle interrupts laser beams to calculate its length and shape. All the information is sent to the lane controller, which determines the toll.

VIOLATIONS AND MISTAKES

If the lane controller does not find enough money in an account or simply does not receive any data because the car doesn't have a transponder, it issues a command to save the images taken by two cameras of the license plates of the vehicle. (All license plates are photographed but the images are saved only if there is a problem.) The digitized video images are sent to the customer service center, and a toll violation notice is mailed out. Sometimes the transponder fails to send a signal. Such glitches are usually caused by a transponder that is not mounted in the right place in the vehicle.

OVERHEAD LASER PROFILER

DRIVER FEEDBACK SIGN

CAMERA

EXIT LOOP

TOLL BOOTH

CAMERA   TREADLE

ENTRY LOOP

LANE CONTROLLER

A.V.I. READER

A.V.I. ANTENNA

E-Z PASS LANE

E-Z PASS TAG (TRANSPONDER)

Source: Mark IV Industries, Traffic Technologies

Mika Gröndahl; Photograph by Tony Cenicola

With Electronic Tolls, No Need to Roll Down the Window

By David Kushner

When science-fiction writers conjure up flying cars of the future, they don't usually make them stop to pay tolls. In today's world of automated teller machines, electronic commerce and online banking, the idea of tossing quarters into toll-booth baskets seems quaint.

Someday soon, it may be only a memory. Wireless, cash-less toll collection systems are gaining ground across the United States.

In the New York region, the E-Z Pass wireless toll system is used by millions of subscribers to pay tolls on just about every major bridge, tunnel and highway. To participate, drivers open prepaid accounts, then receive special tags for their cars' windshields. When a car with one of these tags cruises through a toll plaza equipped with E-Z Pass, the account information is automatically received and the toll debited. No stopping. No cash. No rolling down the windows.

After some initial start-up problems, the system has been credited with helping to ease congestion at most of the bridge and tunnel crossings in the New York metropolitan area.

The E-Z Pass program first hit the drawing board in 1990 after regional toll authorities met to discuss their high-technology future. Soon afterward, Delaware, Maryland, Massachusetts, New Jersey, New York and Pennsylvania established an interagency group. The states' goal was to develop a system that would allow a driver to travel from, say, Buffalo to Baltimore without ever stopping to pay a toll.

But electronic toll collection has actually been in development for far longer, said Neil Schuster, executive director of the International Bridge Tunnel and Turnpike Association, a group that is based in Washington and represents public and private toll authorities in 20 countries. In the United States, Louisiana, Oklahoma and Texas were some of the first states to adopt such programs when they installed automatic systems 10 years ago. Today, Mr. Schuster said, about 20 of the 55 toll authorities in the United States have some kind of electronic system in place.

Convenience is not the only goal. The E-Z Pass interagency group hopes that a reduction in toll booth bottlenecks will reduce stress for drivers and help the environment by cutting down on the amount of time cars spend idling, generating noxious fumes. It also estimates that an electronic system can process up to three times as many cars as a human toll-booth worker.

But, said Rena Barta, program director for the interagency group, "no one has lost a job over this." Computerized systems inevitably create new jobs, she said, and many veteran employees have been retrained to service the E-Z Pass system.

Mr. Schuster, however, said that on an international level, it was inevitable the machines would win in the long run. "There are workers who are going to be potentially displaced by this," he said.

Another issue related to the growth of electronic toll collection is safety. The E-Z Pass system can collect information while a car is moving, which is why it increases traffic flow. But most toll plazas were designed with manual collection in mind, so it can be a safety hazard to have some cars stopping for tolls while others are moving right through. In a more technologically advanced toll system, like one in Norway, Mr. Schuster said, that problem is alleviated because there are no booths at all — electronic toll systems monitor drivers across open highways.

Electronic collection has raised some concerns about privacy. In the E-Z Pass system, transponders record the time and location each time a toll is collected. Ms. Barta said this information was not publicized. It is used only in customer service centers for reference, she said. The information can also be subpoenaed by the police.

Mr. Schuster compared electronic toll logs with the record of calls on a telephone bill. "Most people aren't concerned about phone companies misusing their data," he said. "There's no reason to be concerned about this either."

It's probably just a matter of time before electronic tolls become as commonplace as automated teller machines or computer-coded subway and bus passes. Such digital systems point to a day when seemingly disparate financial services will converge. Mr. Schuster said electronic toll transponders in some states, like Texas, could be used to pay for airport parking fees. Could drive-through burgers be next?

Seeing Green

New York City uses computers to control traffic lights at thousands of intersections. Here's a look at the most important elements of the system.

At the control center

WORK STATIONS allow engineers to adjust traffic signal patterns, based on input from the field.

SUPERVISORY COMPUTER has a central database that exchanges information with the area computers. Engineers can view the entire network using this computer.

Each of the 15 **AREA COMPUTERS** controls between 640 and 720 intersections. These computers also monitor sensor data collected at intersections.

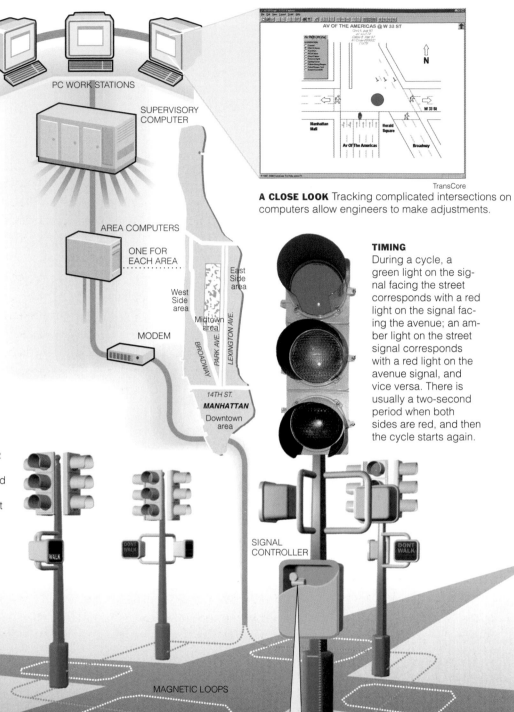

PC WORK STATIONS

SUPERVISORY COMPUTER

AREA COMPUTERS
ONE FOR EACH AREA

MODEM

East Side area
West Side area
Midtown area
BROADWAY
PARK AVE.
LEXINGTON AVE.
14TH ST.
MANHATTAN
Downtown area

SIGNAL CONTROLLER

MAGNETIC LOOPS

SCHEMATIC DIAGRAM

AV OF THE AMERICAS @ W 33 ST

Manhattan Mall
Herald Square
Av Of The Americas
Broadway
W 33 St.

TransCore

A CLOSE LOOK Tracking complicated intersections on computers allow engineers to make adjustments.

TIMING
During a cycle, a green light on the signal facing the street corresponds with a red light on the signal facing the avenue; an amber light on the street signal corresponds with a red light on the avenue signal, and vice versa. There is usually a two-second period when both sides are red, and then the cycle starts again.

At the intersection

Every **SIGNAL CONTROLLER BOX** has its own address; when information addressed to a box is transmitted, the signal controller adopts that pattern at its intersection. Default timers control the 4,800 non-networked signal controllers. If a networked signal controller loses its link to the traffic center, it will automatically revert to a default timer.

MAGNETIC LOOPS embedded in the road at major intersections detect metal from passing cars and report the traffic counts back to the control center.

WALK

DON'T WALK

DON'T WALK

Inside the Signal Box

At the intersection, lights are controlled by electro-mechanical switches. Inside each signal controller box is a constantly rotating dial and a long bar with up to 27 cams (depending on how many lights are at the intersection) that control the different light phases. Each circular cam has ridges on it that activate one of three phases for a traffic light.

Each full circle of the dial corresponds to a full cycle of traffic lights. The cycle length can be set to 60, 90 or 120 seconds.

Each time a timing key reaches the top of the dial, it connects the power that advances the cams one step, changing the color of the signal lights.

Source: New York City Department of Transportation

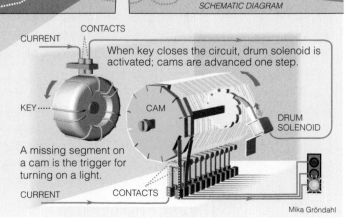

CONTACTS
CURRENT

When key closes the circuit, drum solenoid is activated; cams are advanced one step.

KEY
CAM
DRUM SOLENOID

A missing segment on a cam is the trigger for turning on a light.

CURRENT
CONTACTS

Mika Gröndahl

Choreographing the Dance of Traffic Lights

By Catherine Greenman

In commuter heaven, all traffic lights would magically switch to green for approaching drivers, forming strings of fiery green balls that went on for miles. Pedestrians in paradise would encounter "Walk" signs permanently aglow, and drivers would wait patiently as the walkers dawdled the day away crossing the street.

In real life though, drivers inevitably hit red. And in most cities, when cars must stop is usually controlled by a computer network. In New York City, the network that controls many of the city's lights is based in a single building in Queens, at the Traffic Management Center of the city's Department of Transportation.

How does an engineer at the traffic center tell traffic lights miles away what color lights to flash, and for how long? The center's computers connect to Manhattan traffic lights by sending data through large coaxial cables to signal controller boxes at intersections (the other boroughs are connected via phone lines).

Of the 45,000 intersections in the five boroughs, 10,800 are directed by traffic signals. The center's computers control 6,000 of the intersections via signal controller boxes; each box is wired to all the traffic lights at that intersection. The other 4,800 traffic lights run on timers with preset patterns. If people report a problem with one of these non-networked signals, a traffic center engineer will go out and adjust it manually.

The city is divided into 15 clusters, or neighborhoods; each has its own computer and has 640 to 720 intersections. Manhattan has separate clusters for the East Side, the West Side, uptown, midtown and downtown. Fifteen area computers at the traffic center send information to each cluster of intersections, and these computers are in turn networked to a supervisory computer, which monitors the entire system and coordinates traffic patterns among them.

Every morning, traffic center engineers sit in front of computer terminals and train their eyes on a wall of monitors that display surveillance gathered by 55 closed-circuit cameras located at major traffic arteries. The buzz of a radio stream coming in from police officers on traffic patrols is interrupted several times an hour by the shrill ring of a telephone as one of the several hundred officers or a traffic center official calls in to report a problem in the field. Using that information, the traffic center can respond to different situations and adjust the light patterns.

Traffic lights run on timed cycles that last 60, 90 or 120 seconds. In general, green lights stay on longer for traffic on the north-south streets in Manhattan and on major roadways in all of the boroughs.

Under ordinary traffic conditions, the area computers controlling each cluster of intersections signal a chain of lights to turn green in 10-second progressions. That gives each light you're driving toward time to turn green if you are driving about 30 miles an hour (which is the New York City speed limit). Under that pattern, you should hit a long series of green lights on major arteries before having to stop.

The traffic center uses the same principle — but with red lights — on major arteries to divert traffic away from problem areas. If an accident occurs in Times Square, for instance, the center can create a red-light pattern on nearby streets that make it preferable to find alternative routes.

City traffic would be much worse if the traffic center did not intervene, said Jack Larson, deputy commissioner for operations at the Department of Transportation. "When a major north-south artery in the city is blocked, and the timing isn't modified, the backup just gets longer and longer," he said. By holding more traffic back with red lights at a significant distance from an incident, the traffic center gives drivers enough time to go down a side street before getting stuck at the site of the incident.

The traffic center engineers also create patterns via computer to favor traffic on the roads leading into Manhattan during the morning rush period and out of the borough in the evening rush. Other patterns have been designed to accommodate things like Presidential visits, parades, water main breaks and road construction. "Every year, we learn from experience and create new patterns to smooth out the problems," Mr. Larson said.

Holding Pattern

Los Angeles County's transportation authority has been able to shave 25 percent off the travel time on two express bus routes through several innovations, including the use of an electronic system that can keep a traffic light green long enough for a bus to make it through an intersection.

1 A small transmitter on the underside of each express bus emits a low-power signal that is picked up by a loop antenna embedded in the street as the bus passes over it. The signal identifies the bus to the system.

2 Several hundred feet farther, the bus passes over a second antenna loop that is 200 feet from an intersection. The system compares the time this signal was received with the time of the first one to calculate the vehicle's speed.

3 The system determines whether the bus will get through if the light stays green longer or is switched to green earlier (the extra time can't exceed 10 seconds). It sends a signal to the box at the intersection that controls the traffic lights.

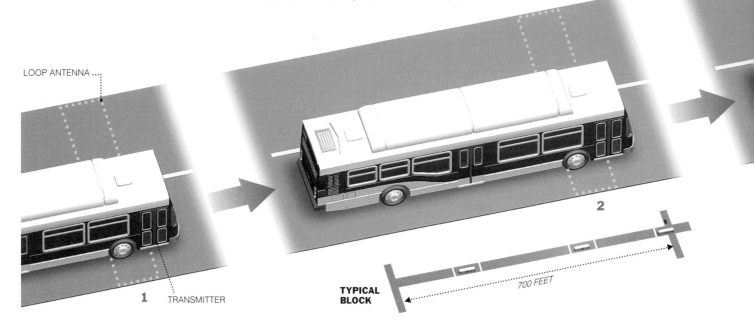

LOOP ANTENNA

1 TRANSMITTER

TYPICAL BLOCK

700 FEET

With Help from Computers, Express Buses Just Keep Rolling Along

By Eric A. Taub

As the growing number of people on the roads in the Los Angeles area clogs traffic, often to glacial speeds, a form of transportation that is invisible to most members of the middle class in that car-oriented megalopolis is helping people travel faster: the bus.

Los Angeles County transportation officials have managed to shave as much as 25 percent off the travel time of some local bus trips by adopting technology that, among other things, can keep green lights on just a little longer as the bus approaches — as long as doing that does not cause another set of traffic problems.

The Metropolitan Transportation Authority learned through surveys that buses spent half their time standing still, either at red lights or at bus stops, waiting for passengers to get on and off. In the late 1980's, officials had tried to speed things

by equipping buses with special transmitters that would hold traffic lights on green until buses passed through. But that just backed up the traffic on cross-streets, so the initiative was abandoned.

In 1997, traffic officials heard about a similar effort in Curitaba, Brazil, that was successful because the system was smarter in several ways. For one thing, it held the lights for buses only when that did not cause other traffic snarls.

Still, there was no guarantee that the system would work in Los Angeles. Curitaba was built to handle public transportation, while public transit in Los Angeles often appears to be an afterthought. Even though one million people ride buses each day, the bus is the transportation mode of last resort for most Angelenos.

The regional transit authority decided to adopt many of

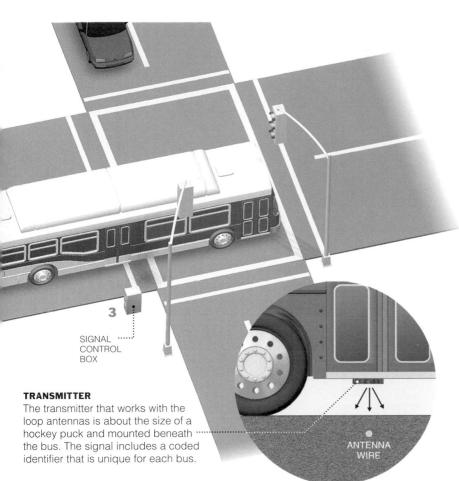

3

SIGNAL
CONTROL
BOX

TRANSMITTER
The transmitter that works with the loop antennas is about the size of a hockey puck and mounted beneath the bus. The signal includes a coded identifier that is unique for each bus.

ANTENNA
WIRE

Keeping track

Signals from the antennas are used to monitor each bus's progress, which is displayed on computer screens at the transportation agency's control center. If a bus catches up with one ahead of it, its icon glows red, and the driver of the trailing bus is told via radio to slow down.

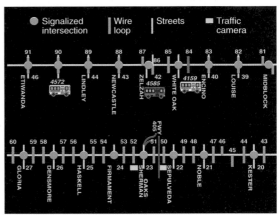

Other keys to speed

Passengers can get on and off the express buses quickly because the buses are low to the ground. Large exit signs encourage departing passengers to leave through the rear doors, and that also helps people board faster. Express routes have been streamlined so buses have fewer stops to make.

Sources: Los Angeles County Metropolitan Transit Authority; North American Bus Industries Inc.

Frank O'Connell

Curitaba's strategies. "Our goal was to decrease travel time over our existing limited-stop express service by 25 percent," said Rex Gephart, who manages the Metro Rapid project. "If we couldn't do that, we'd abandon the project."

The transit authority first had to embed 210 antenna loops in the pavement at various spots along two routes. As a bus passes over one of the loops, a $75 transmitter mounted on the front of the bus sends an identifying signal to an equipment box that controls the traffic light at the next intersection. The signal is also sent to a central control center downtown, so the bus can be tracked in the computer system.

But the project needed to find a way to ease the way for buses to clear intersections without tying up traffic on cross-streets. So the Los Angeles Department of Transportation wrote software that lets a green light be extended — held on green longer or switched to green earlier — for no more than 10 seconds. If several buses approach an intersection as the light is about to change, they can still get only 10 more seconds of green. Buses arriving later than that have to wait. And at important intersections, the green light can be extended in only every other cycle.

To shed additional seconds off travel time, the two bus routes were simplified and stops were limited to no closer together than every eight-tenths of a mile; on local routes, stops can be two-tenths of a mile apart. The express buses also stop on the far corners of intersections, after traffic lights. Stops for local buses are still on the near corner, so the stops are less crowded, making it easier to board and unload passengers.

The movement of each express bus is tracked in the authority's bus control center downtown, both on a computer screen (using the transmitter signals) and through information from video cameras placed at strategic intersections throughout the region.

As a bus passes over a pair of electronic loops embedded in the street, its speed is calculated. Then its arrival time is transmitted via a cell phone link to an electronic display at the next bus stop. Buses are dispatched every 3 to 10 minutes. And if a Metro Rapid bus finishes its route quicker than scheduled, that's fine. That just makes for a more contented rider.

Persistent Memory

Commercial aircraft must carry devices that record performance data and pilots' conversations to provide clues to investigators in the event of an accident. Most flight data and cockpit voice recorders use tape loops to record the information; newer recorders, like the one shown here, use banks of memory chips.

DATA STORAGE

The data or voice recordings are stored in digital form, on flash memory chips inside a hardened shell that is the only part of the recorder designed with crash survivability in mind. Signals from sensors around the aircraft — some analog, others digital — are converted into a single digital stream that is sent to the recorder. With voice recorders, analog signals sent to the unit are converted to digital form for storage.

PROVIDING THE DATA

Air speed is just one of the parameters recorded by flight data recorders. It is measured by a Pitot tube, an L-shaped device usually located near the plane's nose. A Pitot tube (named for a French physicist) actually consists of two tubes, one that is open to the onrushing air, the other that has holes on its sides and reflects the static air pressure. The difference between the two pressures, which is measured by a diaphragm, indicates the velocity of the air. A transducer converts the measurement to an electrical signal that is sent to the recorder.

ONRUSHING AIR

STATIC AIR PRESSURE

PITOT TUBE

AIRSPEED INDICATOR

DATA PROCESSOR

To flight recorder

AIRCRAFT INTERFACE BOARD RECEIVES DATA FROM SENSORS

METAL HOUSING

INSULATION

FLASH MEMORY CHIPS

IN THE TAIL

Recorders are located in the tail section, considered the most crash-resistant part of a plane. Their locations are marked on the exterior of the fuselage.

BEACON

To make them easier to locate underwater, recorders include a sonar beacon, or pinger. On contact with water, the beacon (which operates on battery power) transmits a 37.5 kilohertz sound once a second for up to 30 days.

DATA PORT

A port on the recorder is used to download the information and voice data to investigators' or ground crews' computers.

FLIGHT RECORDER DO NOT OPEN

AQUISITION PROCESSOR BOARD HANDLES DATA FOR STORAGE ON FLASH MEMORY CHIPS

TESTING METHODS

Recorders are typically encased in waterproof crash-resistant steel or titanium boxes that are lined with heat insulation. Manufacturers test them to insure that they will meet the minimum standards for survivability.

IMPACT SHOCK
Subjected to a deceleration of 3,400 G's in 6.5 milliseconds.

PENETRATION RESISTANCE
Is hit by a 500-pound object dropped 10 feet above it, and the point of contact is a quarter-inch hardened steel spike.

FIRE PROTECTION
Intense fuel fire: Subjected to 2,000 degrees Fahrenheit for 60 minutes. Smoldering fire: Subjected to 500 degrees Fahrenheit for 10 hours.

IMMERSION PROTECTION
Deep sea pressure: Immersed in saltwater for 30 days at a pressure equivalent to that at 20,000 feet.

FLUID IMMERSION
Immersed in each of the following fluids for 48 hours: aviation fuel, lubricating oil, hydraulic fluid, toilet flushing fluid and fire extinguishing agents.

Source: L-3 Communications Corporation

Mika Gröndahl and Frank O'Connell

Flight Recorders: The Boxes That Live to Tell the Tale

By Matt Lake

Before the mid-1950's, commercial passenger aircraft looked much like the one that parted Ingrid Bergman and Humphrey Bogart at the end of "Casablanca" — they had propellers on each wing, and they flew at what we would now consider a leisurely pace. It was not until the British company de Havilland leveraged wartime research on jet propulsion into a commercial jet airliner called the Comet that rapid air transport became feasible.

The Comet was able to fly faster and higher than propeller planes, and it was put into service three years before any other company could produce a competing jet. But the Comet suffered several high-profile crashes in 1953 and '54, which gave rise to doubts about the future of jet travel. Before the jet age could begin in earnest, the safety of jet aircraft had to be proved.

The problem was a lack of evidence about what had caused the Comet accidents. With no witnesses, scanty evidence about what had occurred and most of the wreckage lost at sea, the post-crash analyses were sketchy. Eventually, some hard evidence turned up revealing that pressure changes had weakened the window structure — but not before frustrated aeronautical organizations across the world had hypothesized at length on the basis of little hard data.

One of the scientists involved, David Warren, devised a plan that he hoped would lessen such frustration in the future. Dr. Warren, a fuel chemist at the Aeronautical Research Laboratories in Australia, proposed that recording devices be put in the cockpit to capture information. Such devices are now present in all commercial aircraft: the flight data recorder and the cockpit voice recorder.

Based on Dr. Warren's 1954 report, the Australian laboratory produced a prototype called the flight memory. The device recorded four hours of pilot conversation and eight instrument readings by engraving data onto a loop of wire, a design that soon gave way to magnetic tape loops. The concept caught the interest of British aviation authorities in the late 1950's, and aviation agencies across the world had begun to mandate their use by the 1960's.

The devices have changed in a number of ways since then, but the basic idea remains the same. When the recorder reaches the end of the tape loop, it simply records over the older data, providing a constantly updated record for use in analyzing precisely what happened during a flight.

In the 1970's, the de facto standard for storing such data was on tape. In the early 1990's, new designs were introduced using solid-state flash memory; the new designs can store up to 700 different flight parameters, enough to cover the huge number of controls and systems in modern aircraft.

The purpose of flight data and cockpit voice recorders, then and now, is to provide air transport safety agencies with enough information to pin down the reasons for accidents and develop safety procedures to prevent them.

Every airliner is loaded with sensors that measure things like the position of engine controls, the altitude, the amount of thrust for all engines and the temperature and pressure in various parts of the aircraft. It's that kind of information — and a whole lot more — that ends up on modern flight data recorders.

Cockpit voice recorders capture not only voices, but also engine noise, alarms and the sounds of the deployment of landing gear, flaps and so on. The machine noise often proves invaluable in analyzing accidents.

The boxes are positioned in the most crash-resistant part of the airplane, the tail. They are also robust in their own right — the parts where data are stored are usually made of steel or titanium so they can withstand impact. They also contain heat shields made of polymers or heat-resistant wax.

After a crash, the recorders go to the aviation safety authority of the country where the accident took place. There, computers download the information that is stored on the data recorder and use it to create simulations reconstructing the flight. These simulations can include all the cockpit gauges and controls and three-dimensional computer models of the aircraft itself.

But flight data recording is not only for post-disaster analyses. Airlines can use information about specific aircraft to help with maintenance and other issues. A 737 that has been shown by the data recorders to be more fuel-efficient than the norm, for example, can carry less fuel and fly more cheaply.

All This, and the Time of Day, Too

The Swiss watchmaker Breitling manufactures a watch that will transmit an emergency signal in the event of a crash or emergency landing in an isolated area. The watch includes a tiny transmitter and two pull-out antennas; after it is used, it must be rebuilt by the manufacturer.

EMERGENCY WATCH

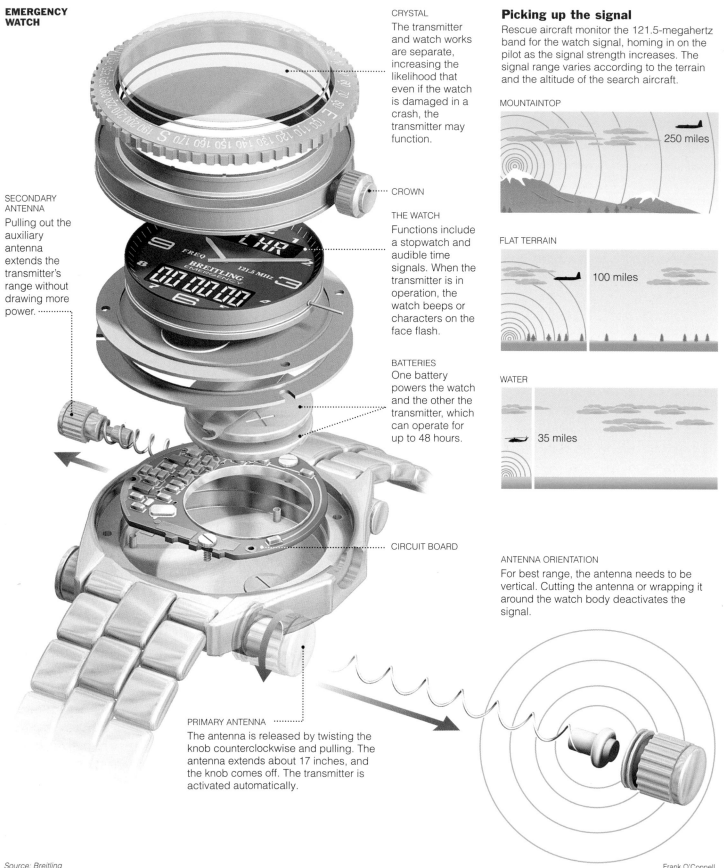

CRYSTAL
The transmitter and watch works are separate, increasing the likelihood that even if the watch is damaged in a crash, the transmitter may function.

CROWN

THE WATCH
Functions include a stopwatch and audible time signals. When the transmitter is in operation, the watch beeps or characters on the face flash.

BATTERIES
One battery powers the watch and the other the transmitter, which can operate for up to 48 hours.

SECONDARY ANTENNA
Pulling out the auxiliary antenna extends the transmitter's range without drawing more power.

CIRCUIT BOARD

PRIMARY ANTENNA
The antenna is released by twisting the knob counterclockwise and pulling. The antenna extends about 17 inches, and the knob comes off. The transmitter is activated automatically.

Picking up the signal

Rescue aircraft monitor the 121.5-megahertz band for the watch signal, homing in on the pilot as the signal strength increases. The signal range varies according to the terrain and the altitude of the search aircraft.

MOUNTAINTOP
250 miles

FLAT TERRAIN
100 miles

WATER
35 miles

ANTENNA ORIENTATION
For best range, the antenna needs to be vertical. Cutting the antenna or wrapping it around the watch body deactivates the signal.

Source: Breitling

Frank O'Connell

On a Pilot's Wrist, Some Life Insurance

By Mindy Sink

Here's a handy accessory for the licensed pilot: a watch that doubles as a rescue beacon. The watch, the Breitling Emergency, looks like an ordinary wristwatch (that is, it looks like an ordinary wristwatch that costs several thousand dollars). But in addition to telling time, it includes a tiny transmitter and two antennas that can broadcast a signal to search-and-rescue crews trying to locate a downed pilot.

"This gives the pilot a personal instrument to wear all the time," said Stefano Albinati, director of the aviation department at Breitling, the watchmaking company in Grenchen, Switzerland. "You may forget a radio or a pen, but certainly you may not forget a watch — especially a pilot."

Breitling has a long history of involvement in aviation, and in 1915 created a wristwatch so pilots would not have to reach into their pockets while flying. The Emergency watch has plenty of features for pilots, including a stopwatch, audible time indicators, an alarm timer and a second-time-zone feature.

But it's when the pilot is in a fix that the $3,500 watch might really prove priceless. By twisting and pulling on a winder cap on the side of the watch — the company recommends that this be done once the pilot is on the ground — the pilot extends a coiled antenna and activates the transmitter, which will broadcast a signal on the 121.5-megahertz band, one of two used throughout the world by rescue units.

In Europe, the watch has been available since 1995 to military and commercial pilots. There has been at least one successful rescue, when a Swiss parachutist became separated from his squadron during a military training maneuver last year and was slightly injured in his jump. He was found relatively easily in remote terrain in a couple of hours.

The company notes that the watch is not a plaything. The transmitter can be used only once, and if it is activated in anything other than a true emergency, Breitling will charge half the sale price to replace the antenna and transmitter. (The company will replace them free if the watch was used in a real emergency.) An unwarranted emergency signal can also mean that the sender will have to pick up the tab for any rescue operation that ensues.

The watch is intended to complement similar emergency locator transmitters that all aircraft, civilian and military, are required to have on board and that are activated automatically in the event of a crash. The on-board transmitter sends a signal that can be picked up by a system of satellites and used to send a rescue crew to the approximate location of a downed plane. Breitling says the signal from the wristwatch should be picked up by receiving equipment on the rescue aircraft, helping the crew spot the pilot and passengers more easily.

The watch is only the latest bit of technology for sending distress signals. From flares and smoke signals, the technology progressed to teletypes and more sophisticated ground-based radio, then to satellite communications. With such technology, said Lt. Cmdr. Paul Steward of the United States Coast Guard in Washington, "you can basically alert somebody on the other side of the world that you're in trouble."

The satellite-based system, known as Cospas-Sarsat (the words combine Russian and English terms for search-and-rescue satellites) was initiated in 1970 after two plane crashes.

"A girl and her mother initially survived a crash, and the girl kept a log of how many planes flew overhead and never saw them," Commander Steward said, adding that the mother and daughter died before the plane was found. Those deaths, and a plane crash in Alaska involving two congressmen, led Congress to require emergency beacons in all aircraft.

These beacons, known as E.L.T.'s, or Emergency Locator Transmitters, are one of three types used with the satellite system. The others are Emergency Position Indicating Response Beacons, or E.P.I.R.B.'s, for maritime use, and Personal Locator Beacons, or P.L.B.'s, which can used only by government or military personnel or in Alaska. They operate on the 121.5-megahertz band as well as on a 406-megahertz band.

The United States is the largest user of Cospas-Sarsat. Commander Steward said there were more than 300 false alarms daily because the 121.5-megahertz frequency could pick up emissions from things like garage door openers and pinball machines as distress signals.

"It's a risk to personnel and a waste of taxpayers' money to respond to these false alarms," he said.

Science
and
Health

All Eyes on Earth

A new generation of satellites is aiming instruments at Earth, giving scientists their clearest picture yet of how the land, oceans and atmosphere interact. The first Earth Observing System satellite, Terra, below, was launched in December 1999. It carries five remote-sensing instruments.

The Measurements

LAND
Surface temperatures, surface features, vegetation patterns

CLOUDS
Extent of cover, structure

OCEANS
Surface temperatures, biological changes

ATMOSPHERE
Greenhouse gases, carbon cycle gases, aerosols

RADIATION
From the Sun, emitted by Earth

SCAN DIRECTION

Terra

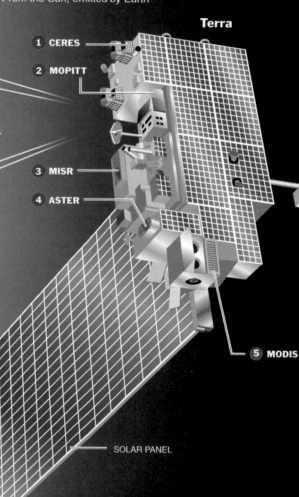

1 CERES
2 MOPITT
3 MISR
4 ASTER
5 MODIS

ANTENNA

SOLAR PANEL

A Leaner Plan

The initial plan was for EOS to consist of six large satellites, each with 12 instruments, but NASA scaled that back. Other satellites in the system are smaller and simpler than Terra. Some make only a single type of measurement.

The Instruments

1 CERES
(Clouds and the Earth's Radiant Energy System) measures incoming solar radiation and radiation re-emitted by Earth, both important in the greenhouse effect.

2 MOPITT
(Measurements of Pollution in the Troposphere) measures gases in the atmosphere to further the understanding of the carbon cycle and the greenhouse effect.

3 MISR
(Multi-Angle Imaging Spectro-radiometer) looks at the structure of clouds and aerosol concentrations from several angles and records things like the heights of trees.

4 ASTER
(Advanced Space-borne Thermal Emission and Reflection Radiometer) produces high-resolution surface images and more precisely analyzes things like clouds and temperature.

5 MODIS
(Moderate Resolution Imaging Spectro-radiometer) makes accurate measurements of global surface temperatures as well as vegetation, ocean and atmospheric changes.

Sources: NASA; Lockheed Martin Astro Space

Frank O'Connell

In Orbit, Remote Sensors Keep Watch Over the Environment

By William K. Stevens

For some of the most sophisticated spacecraft in orbit, the surveillance target is not some distant world, but rather the home planet of those who built it.

This new generation of satellites is called the Earth Observing System, or EOS, part of a long-term effort to subject the interlinked workings of the atmosphere, oceans and land surfaces to detailed scrutiny from space.

The goal of the project, called simply the Earth Science Enterprise, is to fill important gaps in knowledge that prevent scientists from achieving an adequate understanding of the global environment.

One of the main focuses is the Earth's changing climate. Without the information that only satellites can glean, many scientists say, researchers are hobbled in their efforts to predict how the climate will respond to the combination of natural changes and those forced on the planet by human action — emissions of heat-trapping gases from industrial waste, for instance.

EOS has had a long and contentious history. Once attacked as an egregious example of big science gone wild, of distorted scientific priorities and overblown spending, EOS has been sweated down to a cheaper, leaner, more flexible program that is perhaps better attuned to scientists' actual needs.

The flagship of the system, a spacecraft called Terra, is about a third the size of earlier conceptions, partly because of advancing miniaturization of parts. Even so, many experts, including top officials of the National Aeronautics and Space Administration, say the $1 billion, five-ton craft, with five instruments aboard, is still too expensive, too big and too complex. In response, the emphasis of the program has been shifted, with EOS sensing instruments distributed among a number of smaller, cheaper and simpler satellites. That way, if one satellite is lost, the impact on the whole program is lessened.

Terra, which was launched in December 1999 from Vandenberg Air Force Base in California, carries five remote-sensing instruments that scan the Earth continuously. The spacecraft is in a polar orbit, so that it can cover all parts of the Earth.

The instruments go by their acronymic nicknames, and their functions indicate the scientific breadth of the research program. They include:

Modis (for Moderate Resolution Imaging Spectroradiometer). This is the flagship instrument aboard the flagship satellite. It is designed to provide accurate surface temperature measurements of the entire globe; ground-based thermometers leave gaps and are subject to error. Modis also measures biological changes in the ocean, changes in land vegetation, and changes in the properties of clouds and atmospheric aerosols. The latter are tiny droplets of sulfur that reflect sunlight and minute particles that serve as the nuclei for cloud-forming water droplets.

Ceres (Clouds and the Earth's Radiant Energy System) measures incoming solar radiation, which drives the climate system, as well as radiation re-emitted by the Earth, which, when trapped by greenhouse gases, warms the planet. The instrument also measures cloud properties.

Aster (Advanced Spaceborne Thermal Emission and Reflection Radiometer) makes "zoom" images of terrestrial surface features and takes finer readings of surface temperature and cloud cover.

Misr (Multi-angle Imaging Spectroradiometer) is designed to look at the three-dimensional structure of clouds and aerosol concentrations from several angles. It also makes fine-grained images of surface features — for example, the heights of trees.

Mopitt (Measurements of Pollution in the Troposphere) measures atmospheric carbon monoxide, to help better understand the Earth's carbon cycle, and methane, a greenhouse gas.

These instruments can give scientists the clearest picture they have ever had of how the earth-ocean-atmosphere system functions. For instance, two of the instruments measure aerosols and four measure cloud properties. Lack of data about these two features has been the biggest impediment faced by researchers who use computer models to simulate and predict the global climate's behavior. Without a realistic rendition of cloud and aerosol behavior, the models' predictions about any human-induced global warming are uncertain at best.

Two Points of View

During a mission in February 2000, the space shuttle Endeavour bathed Earth with radar waves to gather data for the most complete high-resolution maps of the planet ever obtained. A technique called radar interferometry, in which two radar images are obtained at slightly different locations, generates three-dimensional maps of the surface.

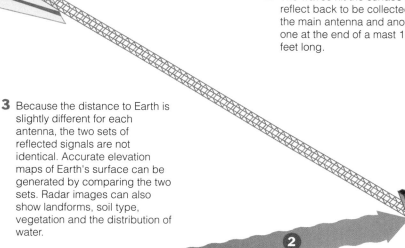

Gathering the data

1 The main antenna located in the payload bay of the space shuttle transmits a beam of radar waves to Earth.

2 The waves hit the surface and reflect back to be collected by the main antenna and another one at the end of a mast 195 feet long.

3 Because the distance to Earth is slightly different for each antenna, the two sets of reflected signals are not identical. Accurate elevation maps of Earth's surface can be generated by comparing the two sets. Radar images can also show landforms, soil type, vegetation and the distribution of water.

The antennas collect data on 139-mile-wide swaths in single passes. The mission surveyed 80 percent of Earth's land.

A 1994 shuttle mission mapped a strip, shown at left, 25 miles wide, in two passes and at a lower resolution.

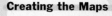

Long Valley region, California

Creating the Maps

The mission is producing topographic maps of Earth far more precise than the best global maps in use today. A key advantage to radar is that it can "see" through clouds and darkness. These three images of the Long Valley region of east-central California illustrate the steps required to produce three-dimensional and topographic maps from radar interferometry.

RADAR IMAGE
This is an image obtained by one antenna. Bright areas are hilly regions with exposed rock and forests. Darker gray areas are sparsely vegetated valleys.

COMPOSITE MAP
When images from the two antennas are combined, the resulting map shows bands of color that represent elevation differences.

TOPO MAP (INSET)
The composite map can then be used to produce topographical maps.

3-D MAP
Composite maps are also combined with computer-generated elevations to create three-dimensional perspectives.

Source: NASA/JPL/Caltech

Juan Velasco

Mapping the Earth, Swath by Swath, From 145 Miles Up

By Warren E. Leary

A project that literally turned the space shuttle upside down is helping to create the most complete three-dimensional map of the Earth ever assembled.

On an 11-day mission in February 2000, the shuttle Endeavour, flying upside down and tail first, with a nearly 200-foot metal and plastic boom protruding from its cargo bay, swept most of the Earth's land masses with radar signals. The Shuttle Radar Topography Mission, as the project was known, amassed mounds of data about the planet's hills, plains, valleys and other surface details.

Comprehensive information about height variations on land aids studies of erosion, flooding, earthquakes, volcanoes, landslides and climate change. The data are also being used to produce improved maps for land and forest management, recreation and the selection of sites for development, including the placement of communications equipment like wireless telephone towers.

Topographical information from ground and air surveys, and from radar data gathered by aircraft and space satellites, is available for most parts of the world in varying degrees of detail. But information gathered at different times from disparate sources is often hard to use in research. Because it is consistent for the whole world, the shuttle data is invaluable.

In addition to topographical maps showing the elevations and depths of the surface, the radar scans produced images of most of the planet's surface area, unobscured by clouds, fog or darkness.

Endeavour, carrying a six-man crew, flew in an orbit that allowed it to map all the land from the southern edge of Greenland to the southern tip of South America, an area encompassing 80 percent of Earth's land mass and 95 percent of its population.

Space shuttles had flown radar-mapping instruments on five previous occasions, culminating in two flights in 1994 that demonstrated three-dimensional mapping by flying over the area twice to take slightly offset images that were later combined into pictures showing elevation differences. Dr. Michael Kobrick of NASA's Jet Propulsion Laboratory, the chief scientist of the project, and another J.P.L. engineer, Edward Caro, came up with the idea of getting three-dimensional topographical images from the shuttle in a single pass by putting additional radar receivers on a boom extending from the spacecraft.

Flying with modifications to the same equipment used in 1994, Endeavour orbited 145 miles above Earth, making radar scans in swaths 140 miles wide. The radar equipment was made by the German Space Agency and the Italian Space Agency.

The shuttle swept the path with C-Band and X-Band signals transmitted by antennas. The signals bounced off Earth and back to the shuttle, where they were picked up by receiving antennas in the cargo bay and at the end of the 640-pound boom.

The radar system uses a technique called interferometry. Each antenna records almost identical data to form images, but the offset distance causes slight phase variations in the signals that allow creation of three-dimensional images when the data is combined.

To a Comet And Back

Interplanetary spacecraft use computers, acting on their own or with instructions radioed from Earth, to stay on their planned flight path. The Stardust spacecraft's maneuvering is particularly intricate: after 7 years it is supposed to return to Earth, with samples of interstellar and cometary dust.

Stardust

ANTENNA

SOLAR PANELS

HEAT SHIELD

PARTICLES

Close Encounter

Comet Wild 2 has made only three passes around the sun since Jupiter's gravity pulled it into its present orbit, so much of its core is intact. Stardust will travel through the coma, collecting dust.

THE PARTICLES
Tiny grains of the comet are remnants of the original dust cloud that coalesced into the solar system.

SAMPLE RETURN CAPSULE
Pictured with its heat shield open, it acts as both container for the collector and its captured samples and re-entry vehicle on the return to Earth.

SAMPLE COLLECTOR
Covered with an ultralight, sticky foam called Aerogel, it will capture thousands of microscopic particles for later study.

SHIELDS
Bumpers protect the craft by deflecting particles away from the main body and solar panels as Stardust nears the comet.

THE COMA
Gas and dust blown from the rocky nucleus form the great glowing cloud that surrounds the head of the comet.

THE NUCLEUS
Roughly four miles wide, the comet's core is composed mainly of water ice, ammonia and rock.

Coming Home

Stardust was launched in 1999 and is due back in 2006. After separating from the main body of the spacecraft, which will remain in orbit around the Sun, the Sample Return Capsule will plunge into Earth's atmosphere at 28,600 m.p.h. It will land in the Utah Test and Training Range, a vast unoccupied salt flat outside Salt Lake City, where its cargo will be retrieved for study.

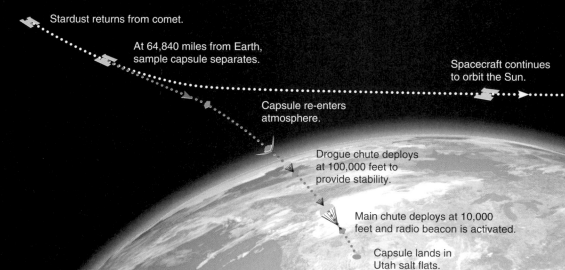

Stardust returns from comet.

At 64,840 miles from Earth, sample capsule separates.

Spacecraft continues to orbit the Sun.

Capsule re-enters atmosphere.

Drogue chute deploys at 100,000 feet to provide stability.

Main chute deploys at 10,000 feet and radio beacon is activated.

Capsule lands in Utah salt flats.

Source: NASA

Frank O'Connell and Jim McManus

In Search of Star Dust and Clues to Life

By William J. Broad

It is the stuff of people, as well as cats, plants, buttons, shoes, seas, planets, comets, moons, cars, books and cell phones, not to mention the paper on which these words appear: star dust.

For decades, science has known of the importance of star dust in the cosmic scheme of things, especially in matter more elaborate than the simple elements made in the primordial Big Bang. But scientists have never had pure samples to study. Soon, though, they will, if everything goes as planned.

A National Aeronautics and Space Administration spacecraft meant to capture interplanetary and interstellar dust in some of its rawest and purest forms is streaking through the solar system. After a seven-year trek, the craft is to return the samples — including many thousands of grains from a comet — to Earth.

Appropriately enough, the mission is called Stardust.

''We want to find out what this stuff is like,'' said Dr. Donald C. Brownlee of the University of Washington, the chief scientist of the mission. ''It's the building block of planets and human bodies.''

Despite its ambitious goals, Stardust is one of NASA's cheaper space probes, costing $166 million. Contributing to the low price is the recycling of spare parts from earlier craft like Voyager, Galileo and Cassini.

Stardust, which was launched from Florida in 1999, uses a particle catcher shaped something like a tennis racket. The catcher is covered on both sides with an extremely low-density foam known as aerogel. This semitransparent material will slow and stop particles without altering them too much. One side will be used when the spacecraft encounters the comet Wild 2; the other, for interstellar dust gathered between Mars and Jupiter.

Wild 2 is a glob of dirty ice a few miles wide. It is considered an ideal target because it only recently has been deflected by Jupiter's gravity from a distant orbit into the inner solar system, so its outer layers have undergone relatively little solar heating.

Wild 2 travels in a looping path from just outside Jupiter's orbit to just inside that of Mars, where it makes its closest approach to the Sun and reaches peak activity.

The spacecraft, 97 days after that peak, is to zoom past the comet. The Sun will be able to coax the ice ball into shedding dense swarms of particles and vapors.

At 13,600 miles per hour, the spacecraft is to swoop through the comet's coma, the globular cloudlike mass that makes up the head, passing within 100 or so miles of the core. It will capture dust and photograph the nucleus. Particles will hit the dust catcher at up to six times the speed of a bullet fired from a high-powered rifle, NASA estimates.

The craft's front end and solar panels are shielded with armor plates to protect Stardust from the storm of icy particles.

Scientists say the particles will range from roughly 100 microns (twice the width of a typical human hair) to less than a micron (one-fiftieth of a human hair). A few, they judge, will be large, and a million or so will be smaller than a micron.

After the collecting is done, in January 2006, the craft's 32-inch-wide return pod is to descend by parachute toward the United States Air Force Test and Training range in Utah. The planned landing site is 100 or so miles southwest of Salt Lake City in the desert.

Then, scientists will begin analyzing its samples, working in particular to establish as firm a baseline as possible for the makeup of cosmic dust. ''It will give us ground truth,'' Dr. Brownlee said.

Stardust's return capsule will bring something else back to Earth as well: a wafer of silicon engraved with the names of 136,000 people who responded via the Internet to a NASA ''send your name to a comet'' campaign. The names were electronically etched and are viewable only with an electron microscope. Another chip, with the names of more than a million members of the public, is attached to the dust collector arm and, like Stardust itself, will just keep drifting through space.

Who Needs Dilithium Crystals?

The experimental spacecraft Deep Space 1, which was launched in 1998, is powered by the National Aeronautics and Space Administration's first operational ion engine. The propulsion system uses just one-tenth the amount of fuel of a chemical rocket, making exploration of the outer solar system more practical. Unlike conventional rockets, which provide a large amount of thrust for a brief period and then coast the rest of the way, an ion drive provides a small amount of thrust continually and builds up momentum over time.

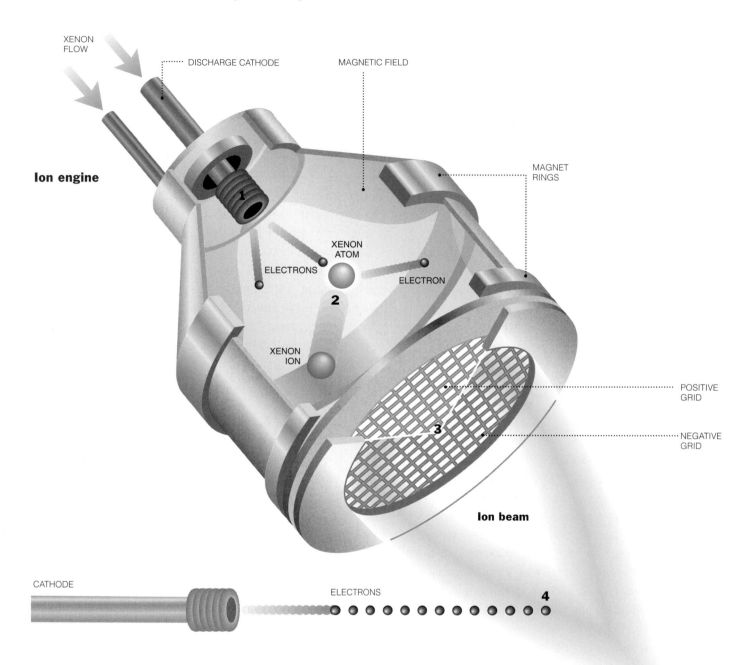

XENON FLOW

DISCHARGE CATHODE

MAGNETIC FIELD

MAGNET RINGS

Ion engine

ELECTRONS

XENON ATOM

ELECTRON

XENON ION

POSITIVE GRID

NEGATIVE GRID

Ion beam

CATHODE

ELECTRONS

A Beam of Particles

1 Negatively charged electrons are generated by a solar-powered cathode, which sits at the edge of a chamber filled with xenon, a heavy inert gas.

2 As the electrons are guided by magnetic fields toward the positively charged walls of the chamber, they strike xenon atoms, knocking off the atoms' electrons. The result: positively charged atoms called ions.

3 As these ions drift toward the open end of the engine, they are drawn to the neutral vacuum of space. Two charged grids focus the exiting ions into a high-velocity beam, producing thrust.

4 To avoid building up a charge in the spacecraft, a second cathode emits electrons into the ion beam, restoring the spacecraft to its neutral state.

Source: NASA/Jet Propulsion Laboratory

Ion Propulsion: Science Fiction Comes to Life

By Warren E. Leary

An experimental spacecraft coursing through the solar system has tested some of the favorite concepts of science fiction. The robot craft, called Deep Space 1, is driven by an advanced ion propulsion engine and takes care of itself, navigating through space on its own and occasionally calling home to let people on Earth know how it is doing.

The 1,000-pound spacecraft, which was launched in October 1998, was a testbed for new technologies that may be incorporated into the next generation of probes sent to explore the solar system and beyond. Although its primary technology-testing mission ended in 1999, it continues to orbit the sun and was scheduled for close encounters with two comets.

Among the technologies aboard are an autonomous navigation system; a miniature integrated camera and imaging spectrometer; an integrated suite of space physics instruments for studying charged particles, or plasma, flowing through space; advanced software with artificial intelligence that lets the spacecraft help run its own mission as situations change, and a variety of new, low-power electronics.

Deep Space 1 has a solar-electric propulsion system, a variation on the concept of ion drive that has been a mainstay of science fiction for decades. While ion propulsion units had been tested in space before, and small ion engines have recently been used to stabilize communications satellites in high Earth orbits, Deep Space 1 was the first spacecraft in which this type of engine was the primary drive system.

"I first heard of ion propulsion in 1968 during a "Star Trek" episode," said Dr. Marc D. Rayman of NASA's Jet Propulsion Laboratory in Pasadena, Calif., the chief mission engineer for Deep Space 1. An ion engine uses fuel 10 times more efficiently than conventional rockets, he said, but it produces an almost imperceptible amount of thrust. This type of engine works by firing continuously for months or years, gradually adding momentum to a spacecraft in the vacuum of space, instead of burning all of its fuel in a few minutes at high thrust, as conventional rockets do.

"Ion propulsion is acceleration with patience," Dr. Rayman said.

At launching, the craft carried 186 pounds of xenon gas aboard to fuel the engine. The xenon, a heavy, inert gas, is not burned like conventional rocket fuel, but charged electrically to create high-speed particles that leave the engine so fast that they produce thrust.

When the ion engine is running, xenon is injected into a chamber and bombarded with high-speed, negatively charged electrons produced by heating a barium-calcium-aluminate material carried in tanks near the engine. These electrons, guided by magnetic rings around the chamber, hit the xenon atoms, knocking away one of the 54 electrons orbiting each atom's nucleus. The resulting atoms, with a net positive charge, are known as ions.

At the end of the chamber, which opens into space, are two metal grids made of molybdenum that are charged positively and negatively, respectively. The charges in the grids create an electrostatic pull on the xenon ions, accelerating them out of the engine and into space at more than 62,000 miles per hour, producing tiny beams of thrust.

The engine on Deep Space 1 produces about 1/250th of a pound of thrust, comparable to the force exerted by a single sheet of typing paper resting on the palm of a person's hand. But this steady thrust increases the speed of the spacecraft by about 30 feet per second each day.

To provide the electrical power to run the engine, Deep Space 1 uses a new type of solar power array never before flown in space that concentrates and focuses sunlight onto 3,600 electricity-producing solar cells. The craft has a pair of solar wings that expand to 38.6 feet across when fully extended and produce 15 to 20 percent more power than similar devices of the same size.

Atop the solar cells sit 720 cylindrical lenses made of silicone. The lenses, which look like glass cylinders cut down the middle, gather incoming sunlight on the rounded outer surfaces and focus it down to the solar cells below the flat sides of the half-cylinders.

Previous unmanned space missions have required ground controllers to constantly monitor conditions on the spacecraft and provide navigational information to keep them on course. Deep Space 1 could keep track of itself, so it required fewer controllers on Earth.

Instead of constantly sending information on its condition, the craft occasionally sends special signal tones that signify different states of well-being. The tones range from "everything is operating acceptably" to "there is a problem that may require help if it worsens" to "need urgent assistance from the ground."

Scanning for Storms

The federal government operates a network of Doppler radar stations to track storms. Like other types of radar, Doppler units send out microwaves in pulses that are reflected back by raindrops or other particles. A computer analyzes the data to determine the distance, direction and extent of storms.

RADOME
Fiberglass shell, 38 feet in diameter, protects the antenna but is invisible to microwaves.

RADAR DISH
Parabolic antenna, 28 feet across, rotates every 20 to 90 seconds, depending upon what kind of weather it is looking at.

FEED HORN
Microwaves, produced at base of tower, travel through wave-guide and out feed horn. Return signal is received here as well.

THE DOPPLER SHIFT
A wave hitting a raindrop moving away from the antenna appears elongated; one hitting an oncoming raindrop appears squeezed. Computers analyze the changes to determine velocity.

ELONGATED WAVE

SQUEEZED WAVE

ANTENNA SUPPORT

ELEVATION CONTROLS
Tilts the antenna up to 20 degrees above the horizon.

COUNTERWEIGHTS

WAVEGUIDE

AZIMUTH DRIVE
Rotates the antenna through 360 degrees.

ACCESS DOOR

MAINTENANCE LADDER

ELECTRIC HEATERS

ELECTRONICS FOR RADAR RECEIVER

TOWER
Typically 15 to 100 feet high.

TRACKING A TORNADO
Radar images, below, from an extremely powerful tornado near Oklahoma City. Top, reflectivity image shows rain intensity; Doppler image at bottom shows wind moving toward and away from antenna (black circle on right).

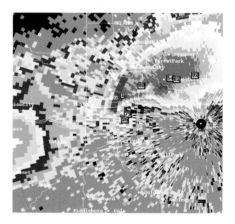

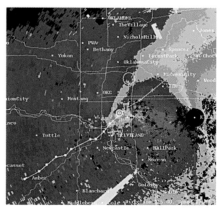

Source: National Oceanic and Atmospheric Administration/ National Severe Storms Laboratory

Frank O'Connell

Now the Weather, Brought to You by Doppler Radar

By Jeffrey Selingo

When severe weather strikes these days, it seems every TV meteorologist can tell you exactly where it's raining or snowing the heaviest, sometimes down to the exact neighborhood and street.

Weather radar has never been as exact or extensive as it is today, said Paul Knight, a meteorologist at Pennsylvania State University. The Doppler radars that many television stations have bought in recent years allow them to pinpoint storms in their broadcast areas. That private network supplements about 120 weather radars operated around the country by the National Weather Service and other federal agencies.

"There are very few places that are not covered by radar," Mr. Knight said.

Weather radar shows where and how much precipitation is falling by emitting microwaves from a rotating antenna. The waves hit anything in their path — like raindrops, snowflakes, hailstones — and are reflected back to the antenna. The radar translates the reflected beams into an image showing the location and intensity of the precipitation, which is displayed through five levels or colors established by the National Weather Service. Green is the weakest; red is the strongest.

But not all radar images are perfect pictures of what is going on in the atmosphere, said Don Burgess, a division chief at the National Severe Storms Laboratory in Norman, Okla.

Like a flashlight beam, waves from a weather radar transmitter expand as they get farther from the source and they do not follow the curvature of the Earth. As a result, radar beams can miss the intensity of precipitation that is far away and close to the ground. Although weather radars have a range of about 250 miles, images within 125 miles are considered the most accurate, Mr. Burgess said.

The type of precipitation also affects radar beams. For instance, light hail may appear as heavy precipitation on a radar screen because it is made of ice and easily reflects radar waves. Snow, on the other hand, does not show up very well on radar images because it is fluffy and usually forms within 10,000 feet of the ground, Mr. Burgess said.

New radar technology has improved the ability of weather forecasters to see snow. In the early 1990's, the National Weather Service began installing Nexrad (for next-generation radar), which was more sensitive and could pick up weaker returns, like those from snowflakes, Mr. Burgess said. "Radars of even the 1980's were pretty crude by today's standards," he said. "Snow was seen so poorly, if at all, in those older weather radars."

The newer Doppler radars are a powerful tool for meteorologists. They allow weather forecasters to determine wind speed and direction, not just the location and intensity of precipitation.

In 1842, Christian Doppler, an Austrian scientist, described how the apparent frequency of waves from a moving object changes depending on whether the object is moving toward or away from an observer. The Doppler effect is why, for example, the pitch of a horn from a car on the opposing lanes of a highway decreases as the car passes by.

Modern weather radar did not come about until a century later, during World War II, when radar was used to track airplanes.

"At that time, weather was an annoying noise that interfered with aircraft tracking," said Greg Forbes, a severe weather expert at the Weather Channel. "Scientists then realized that radars could be used to monitor the weather."

Early weather radars were "like early radios compared to today's sound systems," Mr. Burgess said. Radar screens basically had one color — green — and operators traced weather systems with grease pencils as they passed across the screen. Some of those radars — called the WSR-57 for the year they were designed — remained in operation until 1996, when they were replaced with today's 88-D radars, which refers to the 1988 Doppler design.

Doppler radar research started in the 1960's, but it took several decades for the computer technology needed to run the radars to catch up, Mr. Burgess said. Doppler radar has allowed meteorologists to see certain storms even before they form.

For instance, high winds moving toward the radar, adjacent to high winds moving away from the radar, indicate conditions that can produce a tornado. That has allowed forecasters to increase warning times for tornadoes from less than 5 minutes in the early 1990's to nearly 12 minutes today.

Rough Rider

Researchers and engineers at Pacific Northwest National Laboratory developed sensor fish to help them measure the forces that young salmon encounter when they travel through hydroelectric dams on the Columbia and Snake Rivers.

A trip through the turbines

Fish usually travel through an inlet pipe down through the spinning hydroelectric turbine, then into the draft tube and eventually back into the river. The sensor fish are inserted through a tube directly into the turbine area.

CIRCUIT BOARDS
The fish contain micro-processors and circuitry on their circuit boards to convert the analog signals from their sensors into digital data. They also have memory chips for data storage, output cables for downloading data to a computer and tiny transmitters, which help researchers find the fish more easily once they are through the dam.

AIR BAGS
A gas-creating mixture of chemicals slowly inflates the bags as each fish travels through the dam so that when it reaches the downstream side of the dam, it bobs to the surface for retrieval. Metal rings keep the bags securely attached during the bumpy trip.

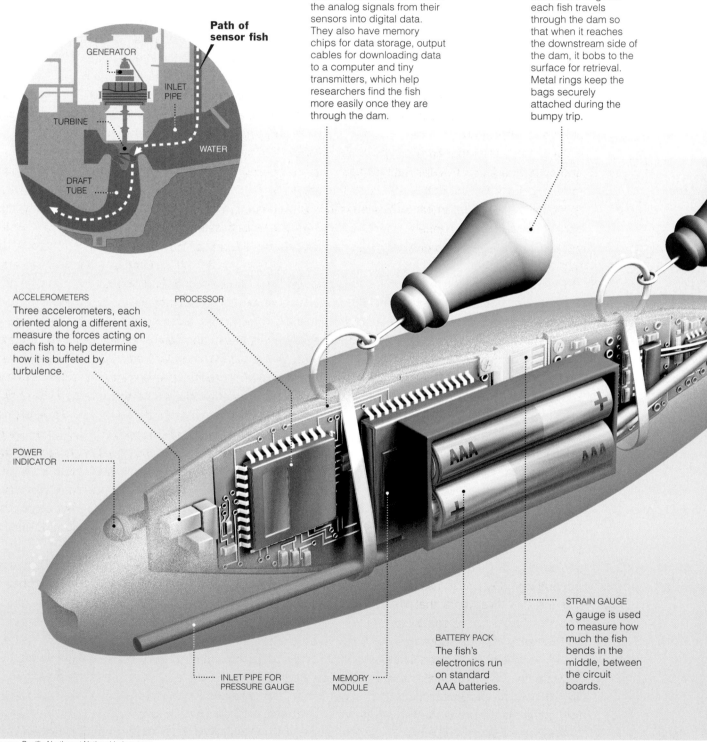

GENERATOR

INLET PIPE

TURBINE

Path of sensor fish

WATER

DRAFT TUBE

ACCELEROMETERS
Three accelerometers, each oriented along a different axis, measure the forces acting on each fish to help determine how it is buffeted by turbulence.

PROCESSOR

POWER INDICATOR

INLET PIPE FOR PRESSURE GAUGE

MEMORY MODULE

BATTERY PACK
The fish's electronics run on standard AAA batteries.

STRAIN GAUGE
A gauge is used to measure how much the fish bends in the middle, between the circuit boards.

Source: Pacific Northwest National Laboratory

Braving the Bumps to Tame Dams for Young Salmon

By Henry Fountain

For a young salmon, spawned on the upper reaches of the Columbia or Snake Rivers in the Pacific Northwest and determined to make it to the Pacific Ocean, a trip through one of the rivers' many hydroelectric dams is like a ride on a bad roller coaster, with a giant blender thrown in for good measure. Severe pressure changes as the fish travels through the dam can leave it stunned and disoriented, easy prey for other fish and birds once it gets through. And the spinning turbine blades that power the electric generators can strike and kill the fish.

Over the years, the Army Corps of Engineers and state and federal wildlife agencies have tried to help juvenile fish avoid this wild ride altogether. Screens have been built near turbine intakes to divert fish into bypass channels. Barges and tanker trucks have been used to give young salmon a lift past the dams.

Despite these efforts, millions of unlucky fish inevitably end up going through the turbines. So some scientists have focused their attention on making the ride a less lethal one. And they've developed a unique tool — what might be called bionic salmon.

These fake fish, developed by the Pacific Northwest National Laboratory, contain tiny sensors and other electronics to measure stress and strain as they flow through the dam (they don't swim). The goal, said Thomas J. Carlson, who manages the project for the laboratory, is to understand the conditions the fish encounter. "If you know those, you can go back and design turbines for safer fish passage," he said.

The idea of creating the fish came up about a decade ago, Dr. Carlson said, "but it's only in the last three or five years that the technology has been available to build something like this."

The fish, which are about six inches long and, in the earliest versions, coated in rubber, include accelerometers that can measure the effects of the turbulence, the pressure changes and the shearing forces that occur as the fish are bounced and buffeted through the turbine and through an outflow tunnel, called a draft tube.

The sensor data are stored in memory chips and downloaded later when the fish are recovered. To aid retrieval, the fish have tiny air bags, which inflate to carry them to the surface, and tiny transmitters that help researchers find them.

The sensor fish have been used at Bonneville Dam, the last one that fish encounter on their way to the sea, to test the effectiveness of a turbine design in protecting fish. Since many fish that are injured in turbines are caught in the gaps between the blades and the turbine walls, the new design has narrower gaps to make it safer for the fish.

The number of fish killed or injured by the blades of any of the turbines along the river is actually quite small, Dr. Carlson said, between 2 percent and 4 percent. Of greater concern is what happens in the turbulent water of the draft tube. Using the sensor fish, the researchers have found that where a fish ends up in the tube is often determined by where it enters the turbine, and that a fish is better off in some parts of the tube than in others.

The sensor fish studies are just a small part of what has been a long and expensive effort to protect salmon and steelhead on the Columbia and Snake, both the juveniles moving downstream to the sea and adults traveling upstream to spawn. For all the billions of dollars spent, the effort has not been particularly successful: some species on the rivers are extinct, and others are endangered.

For some environmentalists and others, the only solution is to demolish four of the Snake River dams. Critics of that plan say that tearing down the dams will not save the fish and will harm the region's economy.

It's not clear to Dr. Carlson that the dams are responsible for most of the problem. "It could be the way that the bypass systems are being operated," he said. Transporting young fish to the ocean on barges, for example, may get them to the ocean too soon and interfere with their development.

What is clear to Dr. Carlson is the need to continue research with tools like the sensor fish so the system can be improved. "The focus is increasingly on fine-tuning elements of this fish-management capability," he said.

The Thoroughly Modern, G.P.S.-Equipped Cow

Scientists in Texas study the grazing patterns of cattle by tracking their paths with the aid of satellites. Each research cow is outfitted with a collar containing a Global Positioning System unit that records the animal's location at specific intervals. The data are downloaded once a week for analysis.

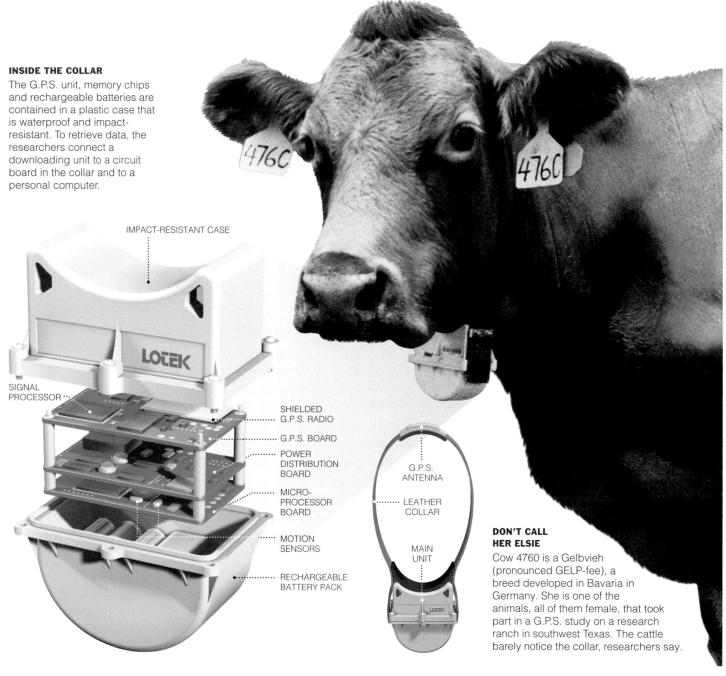

INSIDE THE COLLAR

The G.P.S. unit, memory chips and rechargeable batteries are contained in a plastic case that is waterproof and impact-resistant. To retrieve data, the researchers connect a downloading unit to a circuit board in the collar and to a personal computer.

IMPACT-RESISTANT CASE

LOTEK

SIGNAL PROCESSOR

SHIELDED G.P.S. RADIO

G.P.S. BOARD

POWER DISTRIBUTION BOARD

MICRO-PROCESSOR BOARD

MOTION SENSORS

RECHARGEABLE BATTERY PACK

G.P.S. ANTENNA

LEATHER COLLAR

MAIN UNIT

LOTEK

DON'T CALL HER ELSIE

Cow 4760 is a Gelbvieh (pronounced GELP-fee), a breed developed in Bavaria in Germany. She is one of the animals, all of them female, that took part in a G.P.S. study on a research ranch in southwest Texas. The cattle barely notice the collar, researchers say.

WHERE WAS SHE GRAZING?

The collar, which weighs about a pound, collects several kinds of data.

LATITUDE AND LONGITUDE
The Global Positioning System uses timing signals from satellites to track the cows.

HEAD MOVEMENT
Motion sensors can indicate whether a cow lowered its head to graze or drink, or rested or walked, during a G.P.S. measurement.

ELEVATION
The satellite signals can also be used to determine elevation. A G.P.S. measurement usually takes less than 90 seconds.

TIME
The collar records the time each measurement was taken. That information helps track the route the cow took.

FOLLOW THAT COW

The researchers analyze each cow's G.P.S. data. The data points overlaid on this infrared vegetation map show one cow's wanderings. Her position was checked every 15 minutes. The dots show measurements; the line shows a fence.

Sources: Lotek Wireless; Southwest Texas State University

Frank O'Connell; Photograph by Dr. Lew Hunnicutt

A New Way to Tell When the Cows Come Home

By Shelly Freierman

Dr. Lew Hunnicutt may not know where the buffalo roam, but he has a pretty good handle on the cows.

Dr. Hunnicutt, an assistant professor of range and animal science at Southwest Texas State University, studies the grazing habits of cattle in the hill country between San Antonio and Austin using Global Positioning System technology and digitized mapping. Dr. Hunnicutt hopes to offer new ranchers information on what land to let cattle graze on and what to leave as habitat for wildlife.

Two hundred acres of ranch land does not mean 200 acres of grass, he explained. The brush a rancher sees on the range may hide rocky ground or rich clay-based soil that is fine for grazing. So clearing the brush over some types of rocky ground to grow more prairie grass doesn't make sense if the cattle won't walk over it to feed on prairie grasses.

He and Dr. Bob Lyons, a range extension specialist at Texas A & M University, have conducted a research program that uses G.P.S.-equipped collars developed to track wild animals to pinpoint exactly where cattle are eating or resting on a 4,200-acre research ranch.

The G.P.S. unit records longitude and latitude every 15 minutes with a high level of accuracy. The points of data showing a cow's position are downloaded and overlaid onto two types of maps to figure out where the cows are grazing.

One is a map illustrating soil types, and the other, using infrared images taken from 20,000 feet, shows which parts of the test ranch have thick vegetation and which are rocky.

Dr. Hunnicutt said that maps like those generated in the study can be an educational tool. "It's hard to convince someone not to use 40 percent of their ranch for grazing," he said. "A lot of hill country isn't good for cows because of the rock and brush. The maps can help ranchers visualize the problem."

The G.P.S. collars, which weigh about a pound, are made by Lotek Wireless of Newmarket, Ontario. They are very rugged because they were designed for use with wild animals like elephants, wildebeest, wolves, reindeer, bears and mountain lions, said Leszek Meczarski, the product manager for G.P.S. items at Lotek.

The units are waterproof and can withstand the shock of being banged against rocks or trees by running animals, he said. These G.P.S. devices are one-size-fits-all, and they are fitted onto customized collars for the various animals. Lotek also makes tiny G.P.S. units that are embedded in freshwater and saltwater fish for underwater studies.

In Dr. Hunnicutt's work, each G.P.S. collar, which costs about $5,000, is removed weekly to recharge the batteries and download the data via a cable to a computer. The collars made for tracking wild animals typically have a one-year battery life and a transmitter for sending data to overflying aircraft. Each of those collars costs about $7,000. They also have a release mechanism that can be triggered with a radio signal so the animals do not have to be captured to retrieve their collars.

The cow collar is a vast improvement over the traditional tool for data collection in agricultural studies: graduate students. "We could have used grad students to walk around and follow the cows," Dr. Hunnicutt said. The problem is that when cows see one of the white pickup trucks from the school, they walk toward it hoping for a treat (hay or food pellets), thus changing their behavior.

"With the collars, we can plot their movement without human intervention," Dr. Hunnicutt said.

Still, a graduate student has to go out on horseback twice a day to see if the collared cow is with the herd or off by itself. The student also takes notes on behavior to validate the G.P.S. data.

Up Close and Personal

A swallowable, pill-sized capsule developed by an Israeli company contains a camera, light and transmitter to produce video images of the digestive tract, particularly the small intestine. The images are transmitted to a receiver worn by the patient and are downloaded later to a computer.

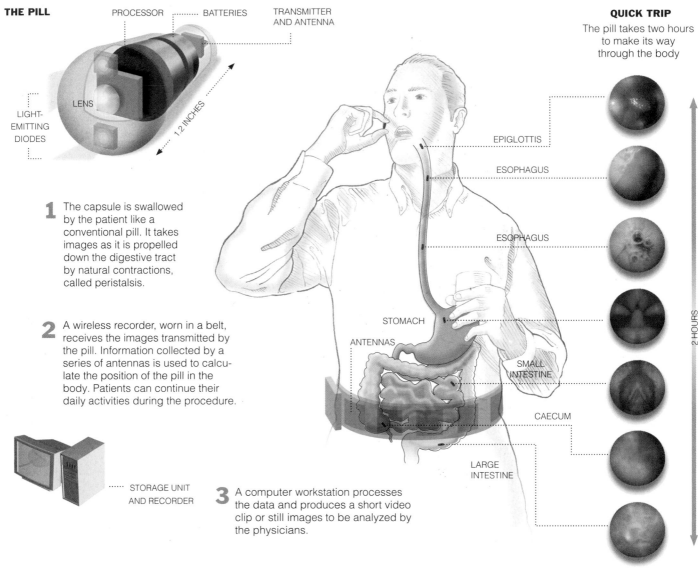

THE PILL

PROCESSOR ⋯⋯ BATTERIES ⋯ TRANSMITTER AND ANTENNA

LIGHT-EMITTING DIODES

LENS

1.2 INCHES

1 The capsule is swallowed by the patient like a conventional pill. It takes images as it is propelled down the digestive tract by natural contractions, called peristalsis.

2 A wireless recorder, worn in a belt, receives the images transmitted by the pill. Information collected by a series of antennas is used to calculate the position of the pill in the body. Patients can continue their daily activities during the procedure.

STORAGE UNIT AND RECORDER

3 A computer workstation processes the data and produces a short video clip or still images to be analyzed by the physicians.

EPIGLOTTIS
ESOPHAGUS
ESOPHAGUS
STOMACH
ANTENNAS
SMALL INTESTINE
CAECUM
LARGE INTESTINE

QUICK TRIP
The pill takes two hours to make its way through the body

2 HOURS

Source: Given Imaging Ltd.

Juan Velasco

"Camera in a Pill" Offers a Travelogue of the Digestive Tract

By Henry Fountain

They haven't figured out how to squeeze Raquel Welch into it, but scientists with an Israeli company have made a pill-size video camera that takes a fantastic voyage of its own, traveling through the digestive tract and transmitting pictures along the way.

The device, which consists of the camera and light source, radio transmitter, and batteries within a sealed capsule slightly more than an inch long and less than half an inch wide, is intended as a less invasive way to examine the small intestine.

Examination of the small intestine, particularly the lower two-thirds, is difficult using fiber-optic or video scopes that have long cables that must be snaked into the digestive tract through the mouth or nose. The procedure usually requires that the patient be sedated.

The camera-in-a-pill "can go some places we have trouble going," said James T. Frakes, a gastroenterologist in Rockford, Ill., and president of the American Society of Gastrointestinal Endoscopy.

The camera takes and transmits several images per second, which are picked up by an array of flexible antennas and a receiver about the size of a personal stereo that are attached to a special belt. The images are stored in memory chips and then downloaded to a computer for viewing, either as still pictures or as a video — the ultimate home movie.

The system uses the strength and phase of the signal received by the antennas to track the pill so that the doctor can know approximately where in the digestive tract a particularly image was taken. So a surgeon, for example, would know where along the small intestine to operate to repair or remove a source of bleeding, a common and often vexing gastrointestinal problem.

The batteries last for about six hours, enough time for the capsule to make it through the small intestine, propelled by the natural contractions known as peristalsis. Patients do not have to interrupt their daily routines during the 24 to 48 hours the pills are in their digestive tracts until excretion.

"Once it's gone down, you don't feel it at all," said Dr. Paul Swain, a gastroenterologist at the Royal London Hospital, who conducted the initial trials of the device. The capsule is designed to be disposable, said Dr. Swain, who has tried it himself.

Dr. Frakes described a demonstration of the device, at a conference in San Diego in 2000, as a "landmark event."

"It demonstrated that you could, in fact, do this," he said. "The possibilities are limitless."

Indeed, the device offers a taste of a future in which molecular-scale sensors, pumps and other devices travel through the body, diagnosing and repairing problems, something like the shrunken submarine in the 1966 film "Fantastic Voyage." The sub, with Ms. Welch and others aboard, was injected into the bloodstream of a gravely ill person, removed a blood clot with a laser and exited through a tear duct (at the last second, in cliffhanger fashion).

The camera-in-a-pill is not quite the stuff of Hollywood dreams, obviously, and one drawback is that it cannot do biopsies, cauterizations or other procedures that a conventional endoscope can.

But Dr. Richard W. Siegel, chairman of the materials science and engineering department at Rensselaer Polytechnic Institute in Troy, N.Y., said the device was "a very nice indication of where things are beginning to go."

"Once you've had success, you're going to want to go on to the next smaller thing, and the next smaller thing," said Dr. Siegel, an expert in the field of nanotechnology, the development of molecular-scale machines.

The new device is a descendant of swallowable capsules with sensors and tiny radio transmitters that have been used for several decades to diagnose stomach problems. Known as Heidelberg capsules, they provide a record of stomach acidity during digestion.

Dr. Arkady Glukhovsky, vice president for research and development with Given Imaging, the maker of the camera-in-a-pill, said the device was made possible by several technological advances, including the development of CMOS video imaging chips. CMOS, or complementary metal-oxide semiconductor, chips, use far less power than more conventional charge-coupled device chips. The capsule also uses newer white light-emitting diodes as its light source. Previously, L.E.D.'s only produced colored light.

The company is working to refine the device and to develop a model that will work in the large intestine and colon, where more light is needed. They also have tested in animals a rudimentary method of propulsion, in which a capsule with tiny electrodes fires a small charge into the intestinal wall, prompting a contraction that pushes the capsule up or down the tract.

While it may be a forerunner of molecular-size devices, this device probably will not get much tinier, Dr. Glukhovsky said. "There's no point to make it smaller," he said. "It should fit the intestine." What is more likely is that the capsule will eventually be able to repair problems as well as view them.

That would reflect another trend in the development of nanotechnology, Dr. Siegel said. "The total package may not want to be small," he said, "but it may be packed with more and more power to do things."

Perhaps, but Dr. Glukhovsky stressed that such multipurpose micromachines were far in the future. As for his company's pill, he said, "If it's a beginning in that direction, it's a very early beginning."

Healing Hands, Made of Steel

The curative ambitions of surgery have always been tempered by the need to cut, penetrate and traumatize the body along the way. Robot-assisted surgery is allowing doctors to get deep inside the body through tiny incisions, reducing the recovery time for procedures as complex as heart surgery.

INSTRUMENTS

An endoscopic camera and instruments like robotic "hands," scissors and scalpels are fitted to the ends of long, pencil-width tubes inserted into the patient.

INSTRUMENT TIPS

THE SURGEON MAKES A KNOT TO CLOSE THE WOUND

TRADITIONAL OPEN HEART SURGERY

INCISION

STERNUM

RIBS

MINIMALLY INVASIVE TELEROBOTIC SURGERY

ACTUAL INCISION SIZE

Gaining access to the heart has typically required surgeons to saw through the sternum and crack open the ribs, leaving a foot-long hole.

Surgeons can now gain the same level of access to the heart by sliding thin, remotely controlled robotic arms between the ribs, through tiny incisions.

ROBOTIC ARMS

A complex structure unifies the operating table and the robotic arms into a single "smart" system that can transmit information to the surgeon's console. For example, the instrument hands inside the body can be equipped with sensors that record subtle changes in tension and resistance. That information is instantly transmitted and replicated in the control instruments held by the surgeon.

SURGEON CONSOLE

The surgeon is able to sit comfortably in front of the main monitoring console, eliminating the physical fatigue and strain associated with hours of standing and bending over the patient.

INSTRUMENTS

Each movement is processed by the system's computer, creating an identical but scaled-down motion at the point of surgery. Advanced systems can eliminate the natural tremor of human hands.

Source: Computer Motion

Tom Zeller and Juan Velasco

Remote Control Surgery: Going Where Fingers Fear to Tread

By Gina Kolata

The heart operation taking place in the pale-green tiled operating room at the Ohio State University Medical Center was almost eerie. The patient, a 62-year-old man, was anesthetized, swathed with blue drapes and lying face up on a narrow table. But no one was touching him.

Instead, the operation was being performed by a robot, whose three metal arms protruded through pencil-sized holes in the man's chest. At the ends of the robot's arms were tiny metal fingers, with rotating wrists, that held a minuscule instrument, a light and a camera. The robot's arms and fingers were controlled by Dr. Randall K. Wolf, sitting at a computer console in a corner of the operating room about 20 feet away.

This sort of operation, heart surgeons say, represents what may become the biggest change in their profession since bypass surgery began in the 1970's. "The reason we make incisions is that we have big hands," said Dr. Wolf, the director of minimally invasive cardiac surgery at Ohio State, in Columbus. The robot's dainty fingers, no longer than a nail on a pinkie, at the end of the long sticks eliminates that constraint.

Eventually, surgeons believe, most heart surgery will be done by robots whose arms are inserted through pencil-sized holes punched in patients' chests. Instead of directly peering into a patient's body, surgeons will view magnified images of the operation on computer screens. In theory, the doctor would not have to be in the same room, or even the same country, as the patient.

Now, a doctor commonly makes a foot-long incision in a patient's chest, saws the breastbone, breaks the ribs to make the holes bigger, then uses a heart-lung machine so the heart can be stopped, often at great risk, for repair.

Robots are not yet good enough to take over most heart surgery. But the promise is that they will enable heart surgeons to fulfill their longstanding dream of less invasive, less traumatic operations. Bypass surgery is indisputably effective, but it is not easy for the patients.

Heart surgeons have spent a decade trying to make bypass surgery less injurious. A few began doing simple cases without heart-lung machines, operating on beating hearts. By pushing down on a part of the heart to hold it steady, doctors can sometimes sew a blood vessel onto a heart that is still beating.

At the same time, doctors had another idea. Why not punch small holes into the chest and insert instruments that are attached to long sticks? They could operate without making large incisions and breaking patients' ribs. But the task proved daunting. This sort of microsurgery, which had proved so successful in other types of operations, was not working in the heart, where precise and delicate movements were impossible with the long sticks.

"It's like holding the end of an 18-inch pencil and trying to write your name," said Dr. Ralph Damiano Jr., who performs robotic surgery at the Milton S. Hershey Medical Center in Hershey Park, Pa..

Robotic surgery was born out of this experience. Several companies worked on methods for doctors to do heart surgery through small holes in the patient's chest, but with a computer controlling the movements of the long sticks.

"It's like taking a computer chip and putting it between a surgeon's hands and the tip of an instrument," said Dr. Michael Mack, director of the Cardiopulmonary Research Science Technology Institute in Dallas. "You can scale up or scale down motion. If it is a 4 to 1 scale, for every four centimeters you move your hands, the tip of the instrument moves one centimeter. You can make a movement as precise as you want it to be."

Those who are testing the robotic systems find their imaginations soaring. Dr. Wolf envisions doing heart surgery on patients in another state or even another country, controlling the robot from as far away as he pleases, while other surgeons stand by. The tiny robot fingers and the precision of its movements could make it ideal for doing heart surgery on fetuses. Dr. Harold Urschel, a surgery professor at the University of Texas Southwestern Medical Center in Dallas, said that the computer, which corrects automatically for tremors in a surgeon's hands, would allow older doctors to operate. "This is going to make you just as good when you're 80 as you were when you were 25," he said.

Sharpening Sight

In 15-minutes, Lasik, or laser in-situ keratomileusis, corrects focusing problems. An estimated 55 million Americans are potential candidates for the procedure, in which a cool laser reshapes the cornea, the transparent layer of tissue in the front of the eye, after a protective flap is lifted.

COMMON FOCUSING PROBLEMS

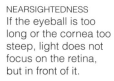
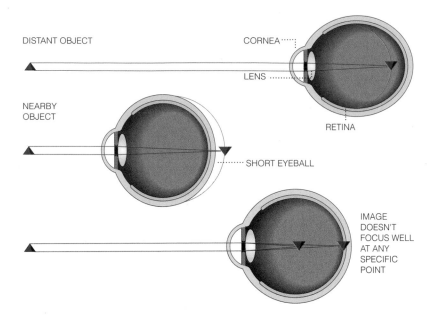

DISTANT OBJECT

CORNEA

LENS

RETINA

NEARBY OBJECT

SHORT EYEBALL

IMAGE DOESN'T FOCUS WELL AT ANY SPECIFIC POINT

NEARSIGHTEDNESS
If the eyeball is too long or the cornea too steep, light does not focus on the retina, but in front of it.

FARSIGHTEDNESS
If the eyeball is too short or the cornea too flat, light from nearby objects focuses in a point beyond the retina.

ASTIGMATISM
An irregularly shaped cornea results in light rays focusing on more than one point on the retina, causing distortion.

Cutting and Reshaping the Cornea

1

OPENING

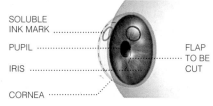

SOLUBLE INK MARK

PUPIL

IRIS

CORNEA

FLAP TO BE CUT

Anesthesic drops are placed on the eye, which is held open. The surgeon marks the cornea with soluble ink to guide how the flap will be repositioned at the end of the surgery.

2

CUTTING

MOTORIZED BLADE

SUCTION RING

A suction ring is used to increase pressure in the eye so that it is firm enough to be cut cleanly. Next a motorized blade, guided along a threaded track, slices the cornea.

3

EXPOSING

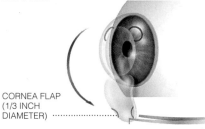

CORNEA FLAP (1/3 INCH DIAMETER)

FORCEPS

The surgeon lifts the flap carefully, exposing the layers of the cornea that will be vaporized by the laser.

4

CORRECTING

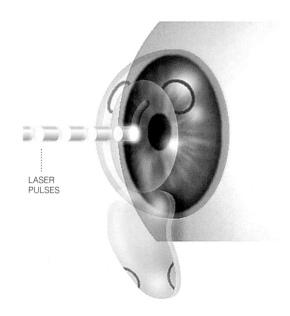

LASER PULSES

NEARSIGHTEDNESS

is corrected by vaporizing the center of the internal cornea to flatten it.

FARSIGHTEDNESS

is fixed by removing a ring of tissue around the center, making the cornea steeper.

ASTIGMATISM

An irregularly shaped cornea is corrected by vaporizing more tissue on one side.

Sources: LASIK Institute; Spectrum Consulting.

Juan Velasco

Laser Pulses Bring an End to Fuzzy Vision

By Kenneth Chang

Harriet Bass lay on a table at the New York Eye and Ear Infirmary, her right eye propped open by a metal ring, staring up at a red dot of light.

"You're going to feel a little pressure," John A. Seedor told Ms. Bass, a 49-year-old casting director, as he was about to slice into her cornea. Using a motorized blade known as a keratome, Dr. Seedor cut a circular flap around the cornea, about one-third of an inch wide and one-150th of an inch thick, and peeled it back.

Switching to a computer-controlled laser, Dr. Seedor trained an invisible beam of cool ultraviolet light on the exposed soft underlayers. The lenses within Ms. Bass's nearsighted eyes bent light too sharply, making her vision fuzzy. The laser vaporized bits of cornea in a way that flattened the curvature at the front of Ms. Bass's eye to bring images at the back of the eye, on the retina, into focus.

"All right, Harriet. It looks beautiful," Dr. Seedor said. He placed the flap back down, caressing it with a brush back into place. "Ready for the other eye?"

Dr. Seedor repeated the procedure. The operation took about 15 minutes.

Although the Food and Drug Administration approved the procedure only in 1999, Lasik, short for laser in-situ keratomileusis, has become very popular, with a million or more Americans a year undergoing the surgery. Intended to correct nearsightedness, the procedure can now treat farsightedness as well.

Lasik's popularity is easily explained: for the vast majority of patients, it is safe, effective and, increasingly, affordable.

"It's a simple procedure," said Richard S. Koplin, medical director of LaserOne in Manhattan. "It's not brain surgery."

With so many people willing to pay out of pocket—most insurance companies do not pay for the procedure, because they regard it as "cosmetic"—to be rid of contacts or glasses, Lasik surgeons have turned to the familiar marketing tools: celebrity testimonials, billboards, radio and television advertisements, telemarketing.

Most Lasik patients encounter no problems at all, although 1 percent to 10 percent require a second Lasik operation to fine-tune their prescriptions.

Still, complications do occasionally occur, some serious.

During surgery, the flap-cutting keratome might jam or create a wrinkle in the cornea, in perhaps 1 in 500 eyes, said Jonathan Carr, director of medical services for the Lasik Vision chain.

After surgery, a range of problems can arise, although most can be easily treated if caught promptly. And, about 1 in 500 will experience an inflammation under the eye, Dr. Carr said. The odds of an infection are about 1 in 10,000. And, there is one reported case of a 35-year-old man who had one eye removed after a Lasik-induced infection.

The operation also cuts down on production of tears, leading to eyes that need drops. Almost always, the condition improves after a few months.

For some, particularly those with large pupils, Lasik can result in nighttime glare and blurry halos around street lamps. Only rarely does bothersome glare persist longer than a few months, Dr. Carr said.

Other surgical alternatives exist, including corrective lenses implanted in the cornea. That procedure costs about the same and is reversible.

And in 2000 the F.D.A. approved a new procedure called laser thermal keratoplasty for treatment of mild to moderate farsightedness in people over 40. The condition affects an estimated 60 million Americans. The procedure takes less than three seconds: the thermal laser fires 16 bursts in two concentric rings, causing cornea tissue to shrink and steepening the edges.

Improvements with Lasik are expected to help extremely nearsighted people and those with oddly shaped corneas. For most, Lasik already produces the results as promised.